A Systems Analysis Model of Urbanization and Change

An Experiment in Interdisciplinary Education

MIT Report No. 20

The MIT Press
Cambridge, Massachusetts,
and London, England

A Systems Analysis Model of Urbanization and Change

An Experiment in Interdisciplinary Education

Carl Steinitz and Peter Rogers

Copyright © 1970 by
Carl Steinitz and Peter Rogers
Set in Linotype Helvetica by P & M Typesetting Inc.
Printed by Halliday Lithograph Corp. and bound in the
United States of America by The Colonial Press Inc.

ISBN 0 262 19070 2 (hardcover)
Library of Congress catalog card number: 70–107996

List of Tables

List of Figures

Participants

Faculty

Carl Steinitz
Associate Professor of Landscape Architecture and Urban Design; Research Associate, Laboratory for Computer Graphics and Spatial Analysis; Harvard University.

Peter Rogers
Associate Professor of Environmental Engineering and City Planning; Research Associate, Center for Population Studies; Harvard University.

Students

Nicholas Dines, Landscape Architecture
Jack Gaffney, Landscape Architecture
David Gates, Landscape Architecture
John Gaudette, Environmental Engineering
Larry Gibson, Landscape Architecture
Peter Jacobs, Landscape Architecture
Larry Lea, Landscape Architecture
Timothy Murray, Landscape Architecture
Harry Parnass, Urban Design
David Parry, Urban Design
David Sinton, City and Regional Planning
Susan Smith, Landscape Architecture
Fritz Stuber, Urban Design
Ghazi Sultan, Urban Design
Thomas Vint, Landscape Architecture
Douglas Way, Landscape Architecture
Bruce White, Landscape Architecture

Visiting Lecturers

Joseph Harrington
Assistant Professor of Environmental Health, School of Public Health, Harvard University.

Henry Jacoby
Assistant Professor of Economics; Director, Harvard Water Program; Harvard University.

The Honorable John M. Quinlan
Senator, 2nd Norfolk District, Massachusetts Senate.

A Systems Analysis Model of Urbanization and Change

An Experiment in Interdisciplinary Education

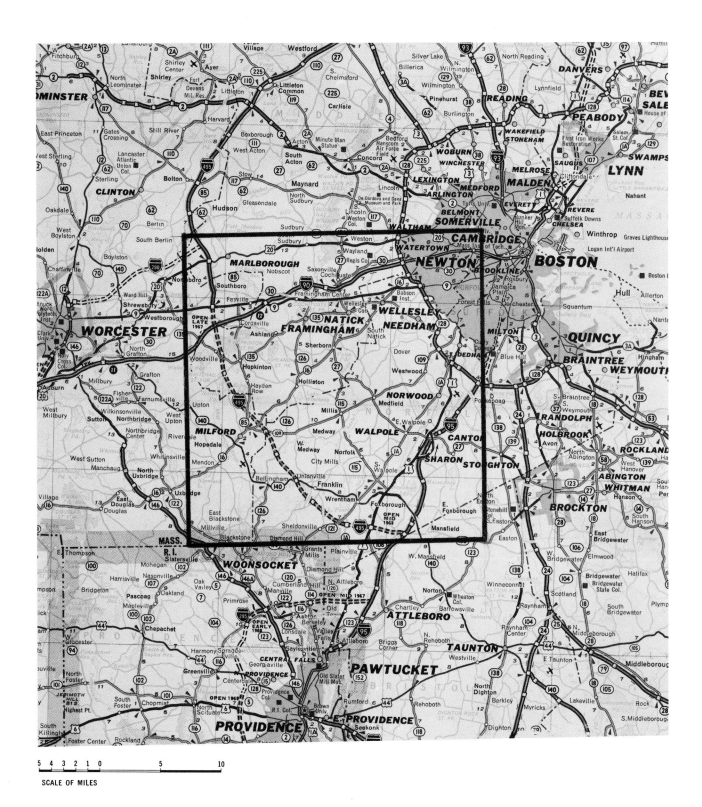

5 4 3 2 1 0   5   10

SCALE OF MILES

Figure 1
Study area.

Introduction

Schools of design have often had interdisciplinary faculties, but to date there has been no system of organization that could bring the various disciplines together into a functioning whole. This book is an outgrowth of an experiment at the Harvard School of Design in synthesizing into a coherent planning model the approaches of four specialties: Landscape Architecture, Engineering, City and Regional Planning, and Urban Design. Specifically, the book reports the activities of a studio course carried out in the spring of 1968, whose study area was the Southwest Sector of the Boston Region, with approximate borders at Framingham, Boston, and Taunton in Massachusetts, and at the Rhode Island state line (Figure 1).

There were two general aims of this experimental studio course. The first was pedagogic: to develop a better method of exploration and of interdisciplinary teaching, which would lead to greater understanding of the complexities of urban development. The second aim, by no means separate or subordinate, was to develop methods that could be used in actual planning and design processes.

Interdisciplinary Education

Interdisciplinary education evokes the idealistic image of synthesizing all available knowledge, techniques, and skills into one supermachine which will solve all problems. However, this interaction often has quite opposite results, such as cross-sterilization rather than cross-fertilization of fields and confrontation rather than dialogue. The communication problems between designers, economists, engineers, planners, and statisticians are immense enough, without adding the necessary psychologists, political scientists, and sociologists. Even the same words may have different meanings in each field. One has only to use the very important word "efficient" to end up with as many meanings as there are disciplines.

Our educational goal necessarily required a loosely structured program, which was frequently modified as the course progressed. The topics were selected according to the interests and capabilities of ourselves and the students. We made no attempt whatsoever to cover the complete range of complexities involved in planning urban development. But we did try to select and model aspects of the environment that we judged to be of basic importance. Because of the varying levels of expertise within the group, some parts of the course were done at a much higher professional level than others. Participants were encouraged to cross fields, so that an engineer might do a visual analysis and a landscape architect a pollution analysis. There frequently was a change of role between teacher and student, when a student happened to be specializing in a particular area relevant to the analysis.

The teachers and the students developed a high tolerance for ambiguity in the course of the seminar. There were times in which the system appeared to break down completely—for example, when nobody knew how to handle a certain problem. But this is, of course, a classroom experience that is pertinent to a real-world situation. Occasions such as this challenged the faculty and students to innovate and to create new approaches.

Often many interests are in conflict in the process of urban development. By assuming these conflicting interests, the group learned to empathize with various points of view, and the process we developed for solving or reconciling these conflicts in our simulation model was perhaps the most educational aspect of the course.

Constructing the Simulation Model

Time was, of course, crucial in the 16-week course, so we obviously had to make many guesses and assumptions; we did not have the time to do the research or to obtain the data that one might want if this were an actual planning problem. Most of the assumptions were recorded as we went along and will be apparent in the presentations.

Model building could begin almost immediately because the basic data had been collected previously in another course in Landscape Architecture taught by Steinitz. The basic spatial unit for data collection and analysis was the one kilometer square Universal Transverse Mercator (UTM) grid. Several of the teams, industry and residence in particular, made allocations on the basis of one-ninth kilometer (approximately 25 acres). We found from aerial photographs that this is a common scale for industry and the usual types of developer housing. The principal data sources were aerial photographs, interpreted with field checks, United States Geological Survey (USGS) maps, reports of Massachusetts state agencies and departments, and the United States Census. Census and other town-based data were linked to the grid through an indexing system. All data were in computer-usable form and were organized for graphic display using the GRID program (Appendix B). The figures were then photographed through a plastic overlay that shows principal rivers and roads.

The participants spent two weeks becoming acquainted with the data; and field trips were made to the study area. During the third and fourth weeks, a series of seminars were held on models as a general field, and specific existing models were examined. The class was guided through selected literature (see bibliography) to determine what sorts of models were feasible to construct and which existing models of urban and regional growth might be utilized. The existing models proved not too useful for the purposes of this course, and more or less original designs were made. The group spent some time on the mathematics of optimization and discussed linear programming and classical optimization; then these were dis-

carded because we were unable at this time to use classical optimizing models for any of our sectorial models. In lieu of this, we used simulation models.

Simulation models are very attractive because of their flexibility, but they are very expensive to perfect. It generally takes considerable time to get even preliminary results, and there was not much computer output after one semester's work. Many of the models for residential growth and urban development reported in the literature have the same limitations as ours.

We relied in the sectorial models upon linear regression models, despite their drawbacks. They are robust and available, and in our simulation model we could use them for residential location, open-space development, and visual analysis. Their drawbacks are indeed serious. They project into the future what happened in the past; they rely on historical data; they do not predict any future change in taste or technology. When looking at our simulation model, however, the reader will notice many "black boxes," representing the component sector models; if we get a better model for one sector, we will pull out the present black box and plug in the new one. This is the simulator's response to attacks on any parts of his model, and the reader will probably notice this several times in the presentation that follows.

During weeks five through eight, student teams prepared models of different sectors of the urbanization process. Each team conducted two seminars, one preliminary and one final, in which their ideas were presented for class discussion. Seminars were also held on the larger issues of urban development design, and the group had the benefit of the special expertise of three guest lecturers: State Senator John M. Quinlin on the political and power structure of his region; Professor Joseph Harrington on pollution; and Professor Henry Jacoby on the question of values and the difficulties from the economic point of view of providing a consistent metric of measurement.

At the end of the eighth week we had, schematically at least, the first articulation of the sectorial models. The faculty then made a first attempt at synthesizing a coherent overall system. As could be expected, all the models were altered in the linking process, since they were constructed without clear knowledge of how they were to be linked. In the subsequent series of sessions on the linked system, it was changed rather substantially, but eventually reverted very nearly to the original suggestion.

Figure 2 shows the simulation model. We began with a projected population increase for our five-year iteration period. We chose four allocation models: an industrial model, a residential model, a recreation and open-space model, and a commercial centers model. We also had a transportation

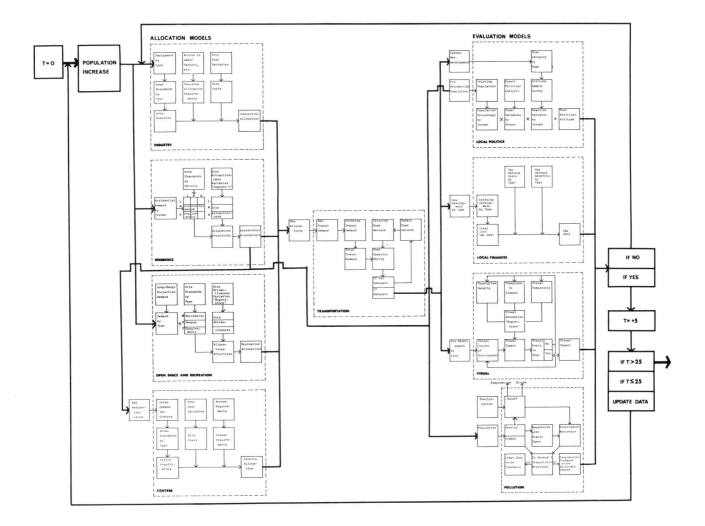

Figure 2
Simulation system.

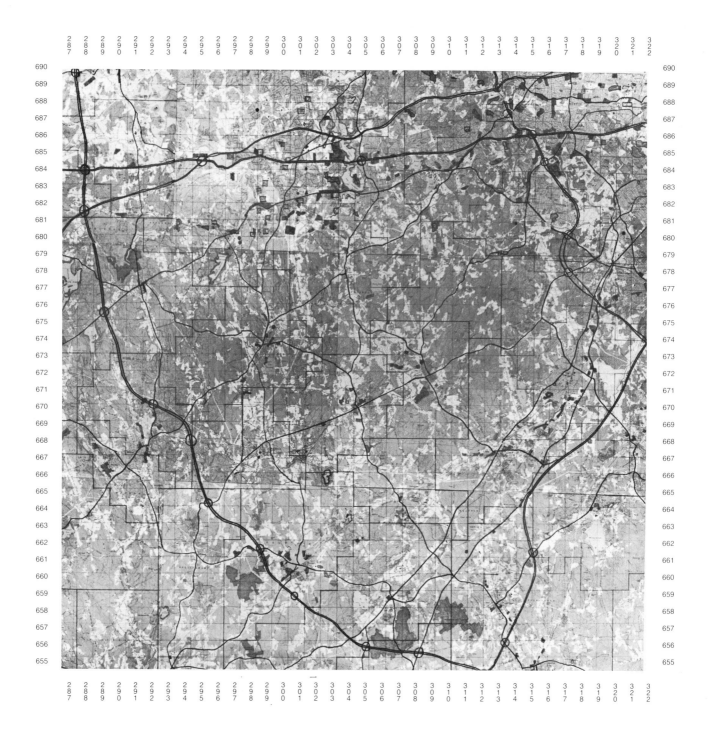

Figure 3
Simulation format.

model, but as it was felt that most of the transportation routes in this region were already established, this model functioned principally as an upgrading process separated from the others. We had four evaluation models: political, fiscal, visual quality, and pollution.

The object of the simulation was to get each of these allocation sectors to make an "optimal" plan within their own objective sets and with respect to their own goals. There was no single objective function to be maximized for the whole region. The sum of each of these sectors was considered the regional optimum. The residential model is essentially a model based on the point of view of maximizing real estate values. In the recreation model, the people whose behavior is modelled are the conservationists and recreationists, and they are maximizing their own objective function. This applied to each of the allocation models. Then each sector was evaluated by the evaluation models. We had many effects between sectors that had to be taken into account when determining adjustments within the individual sectors. That is a very rapid introduction to the flowchart.

It would probably take several years to write a computer program for this simulation. So we did a patchwork process, and dignified it with the name "man-machine interaction." In the process that was finally developed, we used some fully articulated mathematical models. These were developed as a set of rubrics for which the students then played the roles of FORTRAN statements. Of course, FORTRAN is more predictable than students; students in the middle of doing something always ask embarrassing questions. While embarrassing, this is how the model developed. Many of our models were adapted as we went along, and we had to restart several times on the basis of questions which developed while using them.

During the tenth and eleventh weeks, we ran our first trial simulation, linking the data with the models. We operated as though we actually had all available data and a fully operational analysis system. It was during that period that we wrote the final rules and defined the procedures for the simulation process.

The procedures developed for the simulations themselves are analogous to the board, pieces, and rules of Monopoly. USGS maps of the study area were put together on the wall, and the one kilometer square UTM grid was scored on each map. Roads and town boundaries were also drawn. Figure 3 is the resulting board.

Each of the allocation teams was assigned a color and symbol for the land use and its various subcategories, as shown in Figure 4. These were prepared as chips, to be pinned to the USGS maps in the process of allocation. Aluminum-headed pins were used for all new allocations. If a conflict for a cell was present, the chips were pinned on the diagonal. Color-

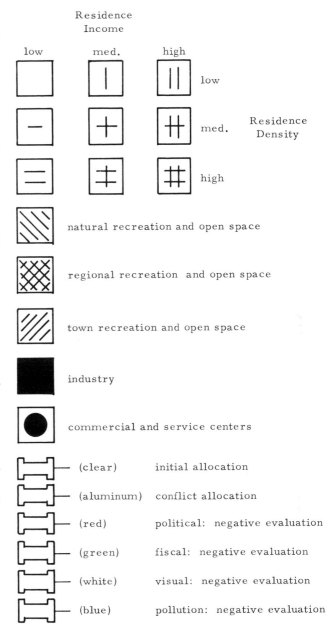

Figure 4
Key to the allocation and evaluation symbols.

coded pins were used for the evaluation models: red for the political, green for the financial, white for the visual, and blue for the pollution models. When an objection was made by one of these models, its pin would be placed on the offending allocation chip. The evaluation pins would be removed as objections were met, either by argument or by reallocation. New locations would be marked with a clear-headed pin and reevaluated. Conflicts among sectors would be similarly resolved and the results evaluated. Finally, with no remaining conflicts or objections, the chips would be stapled to the wall surface and all pins removed. Data files would then be updated, keypunched, and the process would begin again for the next time period (or iteration) in that simulation. The first trial iteration took us two days. The last one, a few weeks later, when we worked out the logistics, took about an hour and a half.

During the following three weeks, we began to run the system model. We did five stages of a first simulation, based on a projection of current trends and attitudes, that is, what would be likely to happen if the region grows but present policies and tastes remain unchanged. This simulation is the base against which all attempts at improvement must be measured. Changes will result in an urban pattern—or a way of life—either better or worse than this one.

We decided that there was time (and energy) left to do one more simulation. The group chose to simulate what might happen if a metropolitan government were established. By doing an alternative with a single, different major assumption the principle of controlled investigation was introduced which could be applied in a third, fourth, or an infinite number of alternatives. Methods of testing change and of comparative evaluation were thus established. As Parkinson would have it, we finished on the last day of the semester.

Systems Analysis and Models for Planning

In order to understand fully our study, it is necessary to have some basic knowledge about systems analysis. This section describes operations research and systems analysis approaches, and in particular the application of this methodology to problems encountered in patterning and guiding the urbanization process. In the discussion that follows, the word *planner* is used in the widest sense, to include all those professionals and roles involved with technological and physical planning. Further, the planner is viewed not only, in the traditional sense, as a designer of specific projects but also as one who translates into public investment programs and planning policies the objectives and criteria of the governmental and economic institutions.

The terms systems analysis and operations research are at present, for all practical purposes, synonymous. Systems analysis is the mathematical analysis of systems of equations; operations research concerns specific decision-making operations. Operations research originated during World War II as the Allies attempted to solve very complex scheduling problems beyond the professional expertise of any one discipline. Interdisciplinary teams were established to synthesize techniques and viewpoints from the several disciplines into a coherent discipline called operations research. The fact that operations research was successful far beyond the expectations of its originators ensured it an important role in the post-war military establishments. The value of systems analysis and operations research for business decision-making was exploited soon after the war, especially by the larger oil and food-processing industries. But operations research techniques were tardily applied in the area of public investment in urban and regional development, as is often the case with technical innovations. This was not due to a lower level of competence on the part of the planners, engineers, and economists, but to inherently more difficult problems.

In discussing systems analysis, we should bear in mind that it is a synthetic discipline; it is a collection of viewpoints and techniques that were formerly parts of such disciplines as mathematical statistics, economics, sociology, biology, physics, and engineering. In the past few years, however, with the increased use of the techniques of systems analysis, a fairly complex and detailed methodology has been derived and accumulated that constitutes a distinct body of knowledge, or at least a distinct way of looking at the solution of complex problems.

Systems analysis could be called a way of looking at problems, and Simon has made the following definition of the new field:

Operations research is loosely defined as the scientific method or straight thinking applied to management problems. This is similar to what had earlier been thought of as "scien-

Table 1
Classification of Models

Type of Model	Example
Iconic	Architect's model of house; engineering drawings; city planners' land-use maps.
Analog	Network flow analyzer using electricity as an analog for water, gas, etc. slide rule addition of logarithms, using length as analog for number.
Symbolic	Mathematical equations, mathematical programming, digital simulation.

tific management" except that operation researchers tend to use rather high powered mathematics. The systems approach is a set of attitudes and a frame of mind rather than a definite and explicit theory. At its vaguest, it means looking at the *whole* problem—again, hardly a novel idea, and not always a very helpful one. The mathematical tools of operations research (including linear programming, dynamic programming, game theory, and probability theory) have a general recipe when applied to management decision making: (1) Construct a mathematical model that satisfies the conditions of the tool to be used and which, at the same time, mirrors the important factors in the management situation to be analyzed. (2) Define the criterion function, the measure that is to be used for comparing the relative merits of various possible courses of action. (3) Obtain empirical estimates of the numerical parameters in the model that specify the particular, concrete situation to which it is to be applied. (4) Carry through the mathematical process of finding the course of action which, for the specified parameter values, maximizes the criterion function.[1]

The most important point of Simon's definition is that systems analysis is novel not merely because it includes more of the system than it had previously but because of its unique overall viewpoint. His definition is, however, rather idealistic in that he talks of maximizing the criterion function. In many practical cases of public investment planning, especially in problems of environmental quality, it is not directly possible to make optimizing models of the system. Under such circumstances, it is necessary to rely on the use of descriptive simulation models instead of analytic models. The distinction between these two types of models will be clarified later.

Types of Models

Since the word *model* is used loosely in everyday language, it is important to define how it is used by systems analysts. Table 1 classifies the types of models, with some examples of their use.

In a systems analysis of a particular problem, we could use any one of the types of models shown in Table 1. However, as experience increases, more and more analysis has moved away from the iconic and the analog toward the symbolic models. In our study of urban growth and change, we, with most other investigators in this field, tended to rely on symbolic models. Lowry[2] classified the symbolic models into (a) descriptive, (b) predictive, and (c) planning models. Descriptive models attempt to replicate the relevant features of an existing investment environment and are useful in formulating theory. The aim of predictive models is to foretell the consequences of an action. Finally there are the planning models, which strive not only to predict but also to evaluate the results in terms of the planner's goals. Lowry listed the

[1]Herbert A. Simon, *The New Science of Management Decision* (New York: Harper and Row, 1960), p. 16.
[2]Ira S. Lowry, "A Short Course in Model Design," *Journal of the American Institute of Planners*, Vol. 31, No. 2 (1965).

essential steps in a planning model as:

1. specification of alternative programs or actions that might be chosen by the planner;

2. prediction of the consequences of choosing each alternative;

3. scoring of these consequences according to a metric of goal achievement;

4. choosing the alternative that yields the highest score.

This is a restatement in more general terms of Simon's definition of operations research.

The symbolic models used in planning analysis can be structurally distinguished as mathematical programming models (*analytic models*) and digital simulation models used for optimization (*simulation models*). By analytic models we mean those based on formal mathematical algorithms.[3] These models are preferred by system analysts because of their mathematical elegance and simplicity. However, they do depend upon the theoretical limits of formal mathematics (usually algebra) and hence are quite often restricted to modelling only the simplest relationships. For instance, nonlinear relationships are difficult to handle by these methods.

An example of a method for analytical model building which has been successfully applied in many cases is linear programming.[4] In this case we are restricted to linear relationships and linear utility and objective functions to be maximized or minimized as the case may be. When the system has been modelled by these linear equations, an algorithm, usually the Simplex algorithm, is then applied to make a solution to the model. Provided that the equations set up as the model of the system meet some minimum requirements, the algorithm will *always* give you the optimal solution (the maximum or the minimum).

Simulation models on the other hand are in general "non-algorithmic." One must create for oneself the logical relationships, which in this case are not restricted to formal mathematics but can include all formal linguistic relationships. It is extremely difficult to construct an optimizing algorithm from such logic. Therefore the thrust of most simulation is descriptive rather than optimizing. This creates some confusion and leads many people to believe that simulation models do not optimize. This is not necessarily the case, since

a descriptive simulation model can be used to make "steepest ascent" approaches to the optimal solution.

An example from transportation planning may help explain how this works. Consider the case of a highway planner wishing to locate a new highway. Since he is also interested in the capacity to which it will be necessary to design the highway, it is not possible to model the problem by an analytical model. There are too many variables that can take on integral values; there are too many nonlinearities; and there are too many time-dependent variables for the present state of mathematical theory to handle. The planner would therefore build a simulation model of the system. To do so, he would assume that he knew where the highway was located and its capacity. He would then write a series of functions that would describe the traffic flow over time (these functions are based on empirically observed phenomena, perhaps coupled with theoretical queuing theory analysis). These enable him to predict how the highway in the proposed location and of the assumed capacity would respond over time both in a physical and economic sense. The planner would then make adjustments to the location, capacities, and so forth, based upon the results of the study, and re-run the model to see if the economic response is better than the first study. In this manner he could improve his design in a series of steps, making an "ascent" on the response surface. As it turns out, there are several different methods for achieving the hopefully-aimed-at "rate of steepest ascent." One such method is the use of repeated random samples of the capacity variables, the so-called Monte Carlo method.

Many types of models, and in particular simulation models, rely on *regression models* as a primary analytic component. They are used in several of the models in our study. Basically, regression is a method to test hypotheses about a particular phenomenon against the observable and measurable effects of the phenomenon. Regression models allow one to make predictions about the behavior of one variable (the "dependent" variable) from observations of the behavior of another variable or group of variables (called the "independent" variables). In strict statistical terms, however, we are not allowed to assume causality between the independent variables and the dependent variable.

For example, you may hypothesize that there is a high degree of association (correlation) between the demand for housing units (X_1) and the size of the population (X_2). You can then make the "simple" linear regression model

$$X_1 = aX_2 + b,$$

where *a* and *b* are constants to be determined by analyzing the recorded data of how X_1 and X_2 vary in relation to each other. There are many statistical tests to show how good this

---

[3]An algorithm is a set of logical rules which enable one to do mathematical operations. For instance, when adding $7 + 14$, one uses an algorithm to obtain 21. Note that one property of an algorithm is that one would always get the same answer: 21 must always be the answer to the addition of 7 and 14. (The algorithm for octal arithmetic would, of course, be different, e.g., $7 + 14$ would be 23.)

[4]"Linear" here means that there are linear proportional relationships between the variables. Take for example, $y = ax_1 + bx_2$. When *a* and *b* are constants, this implies that *y* is a linear function of x_1 and x_2, defined over the entire range of x_1 and x_2.

model is—that is, how good a prediction of X_1 is the variable X_2. One measure is the "coefficient of determination" or R^2. This tells us how much of the variation in the data can be explained by the linear regression model. The value of R^2 varies between 0% and 100%. The higher the R^2, the better the model, is a highly simplified statement about the use of a linear regression model.

Keeping for a moment with the same example, you may feel that, in addition to the size of the population (X_2), you can improve your ability to predict the housing demand (X_1) by adding the per capita income (X_3) into the predicting independent variables. We would now have the "multiple" linear regression model

$$X_1 = aX_2 + bX_3 + c,$$

where a, b, and c are constants evaluated from the recorded data. The values of a and b now differ, of course, from their previous values in the simple regression model. In a similar manner, one may continue to add independent variables. Additional variables, however, may add very little to the ability to predict the dependent variable (X_1). In the examples we report in this book, the regression models were built from the historical data. As the simulation model moved forward in time, the values of the independent variables changed, and from these, assuming that the coefficients a, b, c, etc., did not change, we were able to predict the new levels of the dependent variables.

Urban System Characteristics

From the systems analysis or operations research point of view, there are at least eight characteristics of public investment planning in the urban area which dominate discussion of the subject.

1. *Lumpiness* has to do with indivisibility of projects. Roads, airports, dams, sewage treatment plants, and the like can be constructed only in integral units. It is not possible to talk about one half of one dam or three quarters of an airport or one fifth of one road. Lumpiness causes problems with the mathematical formulation of public investment models, for there are discontinuities in the variables, some of which must be restricted to particular unit sizes.

2. *Economies of scale.* Typically, one expects that as the size of the project is increased from small to large, the economies will be larger. In other words, the marginal costs will become smaller as the size of the project increases. From the mathematical point of view, these economies of scale introduce nonlinearities; and even worse than nonlinearities, they introduce nonconvexity into the formulation of mathematical models.

3. *Long-time horizons.* In any model of the system, we must make projections into the future and the analysis must be carried on in a dynamic rather than a static sense.

4. The *spatial nature* of any investment decision—in urban problems, the specific location—is an important determinant of the investment. A good example of this is route selection for highways in cities. Further, since the spatial demand for goods and services is highly variable over the urban region, there is a proliferation of variables required to model it.

5. *Dispersed decision-making powers.* The simple objective criteria which have been used for models with highly centralized decision-making powers (e.g., in water resources investment) seem no longer applicable in the highly decentralized metropolitan areas. The problem of setting these criteria has not been sufficiently discussed in the literature, and the lack of articulation of these criteria is the weak point of many of the urban planning models to date.

6. The *multisectoral nature* of urban planning admits conflict between sectors, such as housing, highways, or open space. Resolution of these conflicts not only presupposes the objective functions mentioned above, but also the ability to measure the costs, benefits, and qualities of these sectors. With regard to recreation and open space, the models are immediately faced with serious measurement problems.

7. *External economies* (or dis-economies). Obvious examples of this are in the air, water, and solid waste sectors. The intelligent development of sanitary land-fill of solid wastes can have the external effect of producing valuable urban recreation space. There are many good (and bad) spillover effects of investment in public housing. To be able to analyze and capture these external effects, models of the larger system have to be constructed.

8. *Qualitative variables.* These are variables dealing with the quality—or proxies for the quality—of the environment, ranging from "beauty" at one end of the scale to BOD (Biochemical Oxygen Demand) at the other. Besides lacking a convenient market "price" by which they can be added to the monetary costs and benefits of a plan, these variables also generally exhibit large external effects or economies—that is, they have the attributes of *public goods* in that a polluted river affects not only the polluter but everyone else who lives nearby. Examples of such qualitative variables are: (a) visual quality of landscape and townscape; (b) air, water, noise, and solid waste pollution, and (c) ecological and environmental balance.

Urban Planning Models

The final choice of the model to be used depends upon the nature of the systems to be modelled, the use to which the models are to be put, the sort of questions the planning problem poses, and the level of detail required of the analysis. The mathematical techniques available for finding the maximum of a function are the classical techniques of the calculus.

Table 2
Classification of Twenty Urban Planning Models*

Model Name	Author(s)	City	Approx. Date	Land Use a. Residential	b. Industrial (Mfg.)	c. Commercial	d. Govt. or Institutions	e. Roads, Streets, Alleys	f. Public Open Space	Population	Transportation a. Interzonal Trips	b. Other Transp.	Economic activity a. Employment	1. Retail Trade	2. Manufacturing	3. Service	b. Trade 1. Retail	2. Other	c. Personal Income	Projection	Allocation	Derivation	Behavioral a. Economic (Market)	b. Preference	Growth forces a. Gravity	b. Trend	c. Growth Index	d. Input-Output	Econometric and stochastic a. Regression	b. Input-Output	c. Markov Process	Mathematical Programming a. Linear Programming	b. Other Analytic Forms	Simulation a. Autonomous	b. With Intervention
1. How Accessibility Shapes Land Use	Hansen	(Hypothetical)	1959	x						x										x					x									x	
2. Activities Allocation Model	Seidman	Philadelphia	1964	x	x	x		x		x			x		x				x	x	x		x			x		x		x					x	
3. Chicago Area Transportation Model	C.A.T.S. Group	Chicago	1960	x	x	x	x	x	x		x				x					x		x				x									x
4. Connecticut Land Use Model	Voorhees	State of Conn.	1966	x						x			x	x	x				x	x		x			x		x		x					x	
5. Econometric Model of Metro. Employment and Pop. Growth	Niedercorn	(Hypothetical)	1963							x		x	x	x	x				x		x					x		x							
6. Empiric Land Use Model	Brand, Barber, Jacobs	Boston	1966							x			x	x	x				x	x		x			x		x		x						
7. Land Use Plan Design Model	Schlager	S. E. Wisconsin	1965	x		x	x	x	x	x											x	x	x	x								x		x	x
8. Model of Metropolis	Lowry	Pittsburgh	1964	x	x	x				x			x	x							x	x		x							x		x	x
9. A Model for Predicting Traffic Patterns	Bevis	Chicago	1959								x	x								x	x	x			x							x		
10. Opportunity-Accessibility Model for Alloc. Reg. Growth	Lathrop	Buffalo	1965	x						x	x										x	x		x										x	
11. Penn-Jersey Regional Growth Model	Herbert	Philadelphia	196	x																x	x	x	x									x		x	
12. Pittsburgh Urban Renewal Simulation Model	Steger	Pittsburgh	1964	x	x	x				x			x	x				x	x	x	x	x	x		x		x	x				x		x	
13. POLIMETRIC Land Use Forecasting Model	Hill	Boston	1965							x			x	x	x				x	x	x	x			x		x							x	x
14. Probabilistic Model for Residential Growth	Donnelly, Chapin, Weiss	Greensboro	1964	x																x							x		x						x
15. Projection of a Metropolis—New York City	Berman, Chinitz, Hoover	New York City	1960							x			x	x	x	x	x	x	x	x	x										x				
16. RAND Model	RAND Corp.	(Hypothetical)	1962	x						x			x							x	x	x	x	x	x		x		x					x	
17. Retail Market Potential Model	Lakshmanan, Hansen	Baltimore	1964	x							x							x		x	x	x	x		x									x	
18. San Francisco C.R.P. Model	A. D. Little, Inc.	San Francisco	1965	x																x	x	x	x	x								x		x	
19. Simulation Model for Residential Development	Graybeal	(Hypothetical)	1966	x									x							x	x	x	x	x			x							x	x
20. Urban Detroit Area Model	Doxiadis	Detroit Area	1967							x	x									x	x	x			x				x				x		

* Some of the larger simulation models are general systems containing particular submodels of different types, and thus their presentation here may seem to contain contradictions in classification. The apparent contradictions arise from including both general and particular models under one title.

Source: M. Kilbridge, R. O'Block, and P. Teplitz, "A Conceptual Framework for Urban Planning Models," *Management Science*, Vol. 15, No. 6.

With the addition of a constraint upon the function to be maximized, the traditional methods of the calculus have to be modified. In the case where the constraint has to be met with strict equality, the techniques derived by LeGrange in the late eighteenth century may be used. However, if the constraints are inequalities, then the more recent methods of mathematical programming are required.

Since the models that we construct in public investment decision making are not usually of a single dimension, we have to deal with criterion functions of many variables subject to many constraints. In these cases, the mathematical formulation of the model is often quite complex and the simple mathematical and calculus methods referred to above are no longer useful. Often, the major difficulty faced by the systems analyst is to find a suitable mathematical representation of the problem, and it is here that the creative expertise of the analyst is required.

The most comprehensive review of urban planning models is that by Kilbridge, O'Block, and Teplitz.[5] In their paper, they reviewed twenty recent models[6] and classified them according to subject, function, theory, and method (Table 2). Of the studies, only three considered allocation of public open space. Moreover, not one considered the pollution or visual effects of urbanization. None of the models reported a simple overall objective function, in the sense that regional science models usually deal with maximizing G.N.P. or Regional Net Product. In general, the models did not focus on decision making in the sense of classical control theory, in which there are a set of control variables (over which the decision maker or makers have control—budget, zoning, taxes, etc.) and a set of state variables (over which they have no direct control—land uses in the private sector, crime rate, bond rates, etc.). A theoretical model of this kind is given by Weathersby.[7]

It therefore appears that we can report little comprehensive work on the modeling of urban environmental quality. Since most of the models were constructed during the early stages of urban systems modeling, the model protagonists may have decided to "walk before they tried to run." Or, in other words, they chose to consider the more easily quantifiable economic types of variables for their models rather than the more nebulous qualitative variables. No doubt this will change once more experience has been gained with modeling of urban problems.

Of the twenty models classified by Kilbridge et al., only two were analytic mathematical programming models. The remaining eighteen, which incidentally are the more interesting from the point of view of theory and the development of theory, were simulation models. This appears to be the trend in urban planning models.

One of the characteristics of urban planning that we mentioned earlier—that of dispersed decision-making power—is the Achilles heel of the urban systems analysis model builder. In fact, the complexity of modeling the decision-making process is such that one wonders about the advisability of building models at all. There are two answers to this: first, that we are getting better at building models of these types; and, second, that the purpose of model building is educational in a wider sense. We agree with Lowry that,

The participants invariably find their perceptions sharpened, their horizons expanded, their professional skills augmented. The mere necessity of framing questions carefully does much to dispel the fog of sloppy thinking that surrounds our efforts at civic betterment.[8]

In other words, model building produces better planners.

---

[5]M. Kilbridge, R. O'Block, and P. Teplitz, "A Conceptual Framework for Urban Planning Models," *Management Science*, Vol. 15, No. 6.
[6]The dates of the models were from 1959 to 1967. Before 1959, there were no such models.
[7]George Weathersby, "Quantitative Urban Analysis—A Synthesis of Decision Theory and Modern Control Theory," mimeo, Harvard University, Division of Engineering and Applied Physics, Spring 1968.

[8]Ira S. Lowry, "A Short Course in Model Design."

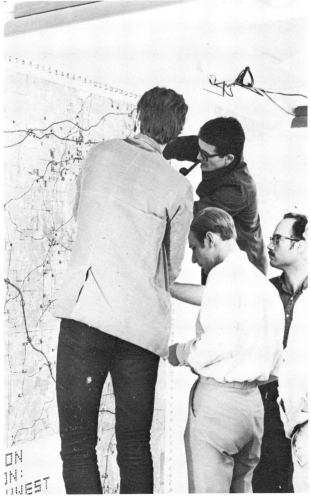

Presentations

The class presented the results of the semester's work at a review attended by representatives of various disciplines both within and without the School of Design. The teams were asked to cover the following points in their presentations: the specific aims that the model in question was addressed to; the key assumptions that they operated under; how the data and the model were then developed to represent the best thinking of the group, which was called the "ideal" state; and how the model changed in use. They were also asked to make an evaluation of how well the model performed—what its limitations and assets were.

Industrial Model (Allocation)

Larry Gibson, David Sinton

Our industrial model represents a sad tale, proceeding from the beautiful to the practical. The model that we started with disappeared almost completely under the requirements of making it operational for the actual simulations. The classical industrial model tries to find an optimum or ideal site for each individual industry. We tried to develop a model to identify sites that would be likely to receive industrial development in the general sense.

Classical industrial location models are primarily concerned with location within large areas where accessibility becomes very important. They give major importance to local conditions about which we knew little—the tax structure and labor conditions. We therefore decided to base our work on a small sample survey of industrial firms in the region, rather than on the studies in the literature.

The first interview was with a major electronics corporation which has just located a very large plant in the town of Hudson. We also interviewed three small firms, two in Southboro and one in Hudson. Several members of the Town of Hudson Industrial Development Commission were also interviewed. As could be expected, the major finding of these interviews was that there are three factors in the location decisions of an industry.

The first factor is the general accessibility of an area. For instance, the electronics corporation, prior to its location in Hudson, had a very good idea of the general area they wished to locate in. The plant that they were going to build must be less than 40 minutes drive from their research plant in Wayland and also be accessible to their potential supply of industrial labor—in this case, Worcester. They virtually defined the general area in which they were going to locate the plant by drawing circles centered on the Wayland research plant and Worcester, and saying that the plant must be within an area defined by the two intersecting areas. This was not an unreasonable first step, but once done, they still had a very large area within which they had to start selecting and evaluating sites.

Their second decision factor was, therefore, to find sites on which they could locate. And this was the problem with which we were most concerned in our development of an industrial location model. There were two different sets of criteria by which sites were evaluated. One set can be referred to as specifically site-oriented criteria, such as the cost of the land, the cost of putting down foundations, the cost of landscaping. We were surprised to find the degree to which the people we interviewed were concerned with their company's image and its site landscaping and whether they fit into the community. And certainly those firms were very much concerned about the costs of landscaping the potential sites. The second set of criteria was costs associated with the municipality or the town within which each site was located: the cost of taxes on their land and their buildings, the availibility and cost of utilities, and so forth.

The third factor in industrial location decisions was difficult to assess—the desire of the town to have the industry, how cooperative the town is, and what sorts of activities are performed by its Industrial Development Commission, if any.

These factors are also found in the classical formulations, with the exception of the great emphasis on site costs. For our ideal model, we developed a cost function for industrial location which took into account transportation costs for labor, materials, and the products of the firms and the site costs of foundations, drainage, and landscaping (Table 3). The model, while valid for individual firm location, was very difficult to generalize. In order to use it, one required very detailed information about each individual industry, as has been indicated in Table 3. Unfortunately, when we made the model operational, we did not in fact allocate industry according to whether it was research, heavy industry, or light industry because we could not obtain a breakdown of which types of industry were likely to locate in the area. This compromised the original model that we produced.

For the simulations, we used a rough generalization of the original model shown in Figure 5, which is based on numerous assumptions. First, we identified those areas which would have relatively low site costs. We looked for cells with low land costs and for cells which were relatively close to major transport routes, and developed a measure of how close each grid cell was to a highway and/or rail line. We eliminated all cells which had very steep slopes, bad drainage conditions, or a heavy bedrock shelf close to the surface. We also took into account the availability of water and sewer facilities. This left us with a number of cells throughout the study area which were relatively inexpensive for industry to build on. The dark cells on Figure 6 show those cells which would probably have low site costs for industrial location.

Table 3
A Model of Industrial Location

Objective function: minimization of the cost of location.

Area or Transportation Costs

(a) Labor. Costs are usually borne by the employee, but if there is no available labor there may be very significant costs involved for the firm. This will include the cost of housing executives. In general, labor is considered in three groups.

L = 1 Managerial or Research
    2 Skilled or Semiskilled
    3 Unskilled

(b) Products and Materials. Both have to be transported to or away from the factory. For the purposes of this study and because we were not concerned with individual industries, we considered industries in the following groups:

I = 1 Heavy industry
    2 Light, transport orientated
    3 Light, not transport orientated
    4 Small industries

For industry type I, the cost function for area location is:

$$(DIST(I) * PIND(I) + \sum_L [DIST(L) \times PLAB(L,I)]) \times N,$$

where
DIST(I) is the distance of a possible site from the transportation node from which the product/material enters the region;
PIND(I) is the price to move one unit of product/material per unit distance;
DIST(L) is the distance from a possible site of the labor type L required for one unit of product;
PLAB(L, I) is the price to move labor type L per unit distance for industry I;
N is the number of units of product.

Site and Community Costs

The cost of land is usually the largest cost associated with a site. However, when two sites with similar land costs are being compared a major determinant between the two will be the marginal costs of such factors as:

J = 1 Foundation Costs
    2 Landscape Costs
    3 Service Costs
    4 Taxes
    5 Industrial Commission Cooperation

It is not easy to compare these costs from site to site on a general basis, so it was decided to compare them to a set of standard conditions which would represent minimum costs.

For industry type I, the cost function for local location is:

$$LC + \sum_J ((STAN(I,J) - EXSIS(I,J)) \times COST(J))$$

where
LC = Land Cost,
STAN(I,J) is standard for factor J,
EXSIS(I,J) is existing state of factor J,
COST(J) is the conversion of the measurement of factor J to dollar costs.

Figure 5
Industrial model (allocation).

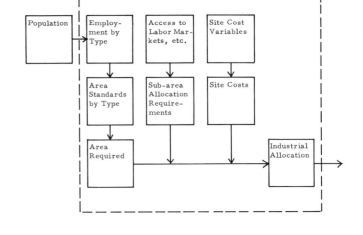

Figure 6
Site costs for industrial development.
. 0 = least
, 1 =
, 2 =
+ 3 =
X 4 =
0 5 =
θ 6 =
θ 7 =
▨ 8 =
▉ 9 = most

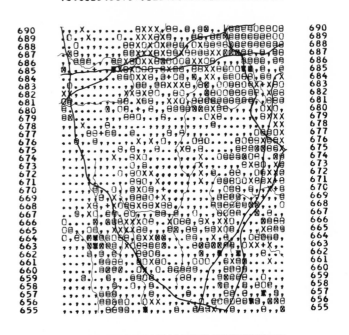

We then had the problem of deciding in which cells we wished to locate. We therefore looked at the access properties of the area. We made a transportation overlay of the area and marked on this map those cells which have good access to the transportation network of larger roads and railways. We felt that location in these would be cheaper than in other equivalent cells because of lower costs of access to major transportation. We had some information on access times throughout the area: from Route 128, from Framingham, and from Providence. Using these access maps, we divided the study area into five subareas which were relatively more accessible to the various external market areas, and we assumed that specific amounts of industry would locate in each. This is a major assumption but it was the best we could do under the circumstances. We did not have the ability really to test accessibility nor did we know to what the industries would want to be accessible. By dividing into subareas, we could ensure that there was a distribution of industry and that the varied accessibility of the market areas was taken into account.

From the population projections that we had, we were able to determine how much industry would be distributed in the area in each time period. We assumed that 30 percent of the population increase in each time period would work within the area; and after studying employment within the suburbs of Boston, we determined that about 38 percent of this increase in the labor force would be working in industry. We distributed industrial employment on the basis of 360 persons working in one ninth of a one kilometer cell. One ninth of a one kilometer cell is 24 acres, and this works out to approximately 15 persons per acre. This is a rather arbitrary figure. In some of the new industrial park developments, the figure is about 7 to 8 persons per acre, while in older areas within the city, the figures can get to over 50 and 60 persons per acre. We took 15 persons per acre as a reasonable figure, though it is perhaps a little high in terms of some modern industries.

From this, we determined how many one-ninth cells we had to distribute, and we then tried to select that many ninths at random from those cells that we had evaluated as being most suitable for industrial development. Unfortunately, we never really got the random selection process operational. We initially tried to use random number tables to obtain a point and then take the nearest suitable cell for industry. This did not work out, since we wanted to get a distribution of industries within subareas in proportion to the relative values of the suitable cells. We therefore used the ancient method of blindfold point selection within subareas and located in the nearest good cell to selected points. This is really a very crude way of selecting the cells. We feel that with more time, we would have worked out a suitable random process for selecting these cells.

For the simulation which was predicated on metropolitan government, we assumed that industry was going to cluster. So instead of looking for individual sites for industry, we looked for sites for large industrial parks; and once we had found them, we clustered the industry within these areas. This was a much less complex procedure. We would look for the good site locations and then place a very heavy emphasis on accessibility to transportation, much heavier than we might normally do. We also allocated a few one-ninth cells on an entirely random basis, on the idea that not all industry is going to cluster.

In the working of this model, we came up against one very basic problem. The average size of an industrial location within this area is one to five acres, but the minimum area we could deal with in the simulation was one-ninth kilometer, or about 24 acres. This meant that we could not allocate industry in the way that it generally allocates itself. This is a problem of the scale of the study, and we think it is something that could be overcome by using a smaller grid.

Another problem we encountered was to what degree these towns or municipalities wanted industry. When we look at the simulation we find that relatively little industry has been put in Wrentham, and we understand from a recent newspaper article that Wrentham desperately wants industry. If we had put this factor into the model, it would have made quite a difference in some of the selected locations, although not very much in the general urban pattern.

In order to make a better industrial model, we ought to have done about seven separate allocations, representing the different industrial types. However, to have tried to allocate seven different types of industry would have involved multiplying the amount of time it took to run the simulations by seven; as time was an all-important factor, we had to compromise by allocating industry in the more general way that we have described.

Residential Model (Allocation)

Jack Gaffney, Harry Parnass, Bruce White

For most of us, the beginning of the semester was the beginning of our experience in dealing with any kind of mathematical model. The very first model that we developed was what Rogers has referred to as a "black box." At that point we had everything imaginable inside the black box because we did not know how much was relevant. As we proceeded, things became clearer, and we constantly altered and updated the model in use. It was not until late in the semester that we developed our model, shown in Figure 7, and produced a

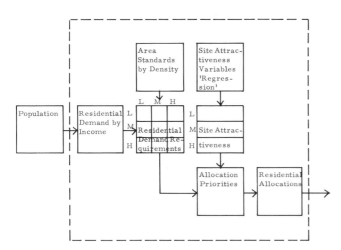

Figure 7
Residential model (allocation).

set of maps that showed meaningful results.

Our ideal model started with the hypothesis that the factors influencing residential location and density could be measured or constrained in such a manner as to allow us to predict where residential development is likely to occur. Before we began we had to have some basis for selecting meaningful figures for residential density. A sketch problem was given in which the whole class developed different types and costs of housing, and we arrived at three density types that we then used to simplify the allocation process. We said that our low density would be 3,600 people per kilometer (15 per acre), that the middle density would be 9,600 people per kilometer (40 per acre), and that the high density would be 19,200 people per kilometer (80 per acre). From the existing data bank on average costs of housing per cell, we assumed that low-cost housing would be under $14,000, middle-cost housing in the range from $14,000 to $25,000, and high-cost housing $25,000 and over.

In order to develop the regression models that were required by our model, we selected 90 sample cells from the data bank on the basis of housing costs: 30 low-cost, 30 middle-cost, and 30 high-cost. We then selected from the data bank those variables which affect housing location and housing density. As the dependent variable in the regression model describing the attractiveness for housing location, we chose land value. As independent variables, we chose the tax rate, access to a limited-access highway, access to Route 128, access to Providence, access to Framingham, distance from an elementary school, the estimation of visual quality, and the educational dollar expenditure per pupil. Based on the data we had to work with, we felt that these were probably the most meaningful factors affecting the location of housing. After performing the necessary computations to develop the regression formulae for low-, medium-, and high-cost housing, we proceeded to develop a regression formula for the factors affecting housing density. In this case, we used the density of the 90 sample cells as the dependent variable. For the independent variables that we felt would affect housing density, we chose topographic slope, the drainage conditions, availability of water supply and, again, tax rate. We would then produce the regression formulae describing low-, medium-, and high-density housing.

We identified the areas that were available for residential development by determining for each cell the percentage which was unavailable—those parts being water, recreation, industry, institutional use, commerce, and existing residential use—and subtracting it from 100 percent. By applying the first regression model to all 1,800 cells we produced three maps of locational attractiveness according to the cost bracket. Using the same process, with our second regression model we would have made a set of maps to indicate which cells were most attractive for the three levels of density.

From this point, we would have liked to have combined the density-potential analyses with the locational-attractiveness analyses to produce a series of maps that would show which cells would be best suited for low-cost and low-density housing, which for high-cost and high-density housing, and which for all of the nine combinations of cost and density. The final step would then have been to distribute housing by cost and density on available land according to market forces.

In use, the model had to be altered. We were never able to get a meaningful description of residential density by housing type as the dependent variable, because the region was essentially all low density. It was difficult to describe medium and high density when the region had insufficient density development upon which to base the analysis.

Before describing the model that finally developed, we will describe some of its assumptions. First, that the measure of the attractiveness of land for housing would equal land value; second, that land costs have a direct correlation to housing costs; and third, that it would be possible to describe the variables controlling these land costs through a regression model.

The first step in the operational procedure was to take the 90 sample cells and select the possible determining variables from the data bank we had for the study area. After selecting the independent variables mentioned earlier, we performed a step-wise multiple linear regression analysis of these variables to determine correlations with the dependent variable which was land cost: in other words, to see if land value could be predicted from the 30 samples for each type of housing and also to determine the regression coefficients of the relevant independent variables by which they should be weighted in the predictive equations. The regression formulae and their respective statistical properties are shown in Table 4.

Subsequent application of these predictive equations to all 1,800 cells and mapping of the resulting analyses with the GRID program produced the three maps in Figures 8, 9, and 10—which show the relative attractiveness of each cell for high-, middle-, and low-cost housing in ten value levels.

As shown in Figure 11, we assumed that 15 percent of this population would be low income, 65 percent middle income, and 20 percent high income. We also assumed that there would be a distribution of demand for the various housing densities. These could be altered in every five-year iteration period. Allocations were made on the assumptions that 63 percent of the total residential development would be low-density, 25 percent middle-density, and 12 percent high-

Table 4

Residential Model: Multiple-Regression Analysis for Low-, Medium-, and High-Income Housing.*

Independent Variable	Low Income		Medium Income		High Income	
	Regression Coefficient	% Increase in R^2	Regression Coefficient	% Increase in R^2	Regression Coefficient	% Increase in R^2
Tax rate	0.032 (.015)‡	5.5				
Access to limited highway					− .824† (.499)	28.0
Access to Route 128	−0.514 (.194)	4.5	−1.705 (.649)†	9.8	−1.774 (.739)	8.5
Access to Providence			−1.134 (.607)	10.3		
Access to Framingham	.502 (.148)	4.3				
Distance to elementary school					1.888† (1.074)	5.9
Dollar expenditure per pupil	.007† (.003)	5.3	.037† (.021)	7.2		
Visual quality			.168 (.061)	5.5	.146† (.075)	5.7
Topography	.494† (.245)	6.0	−1.427 (.519)	13.4		
Drainage	1.172 (.223)	42.0	2.379 (.880)	16.2	2.580 (1.102)	8.5
Intercept	−4.825		−5.671		3.627	
Multiple determination coefficient (R^2)	.679		.622		.568	
$n = 30$	$F = 8.111$		$F = 6.342$		$F = 6.330$	

*Attractiveness measured by use is the dependent variable.
†All coefficients are significant at the .05 level except those marked with a dagger.
‡Figures in parentheses represent the standard errors of the regression coefficients.

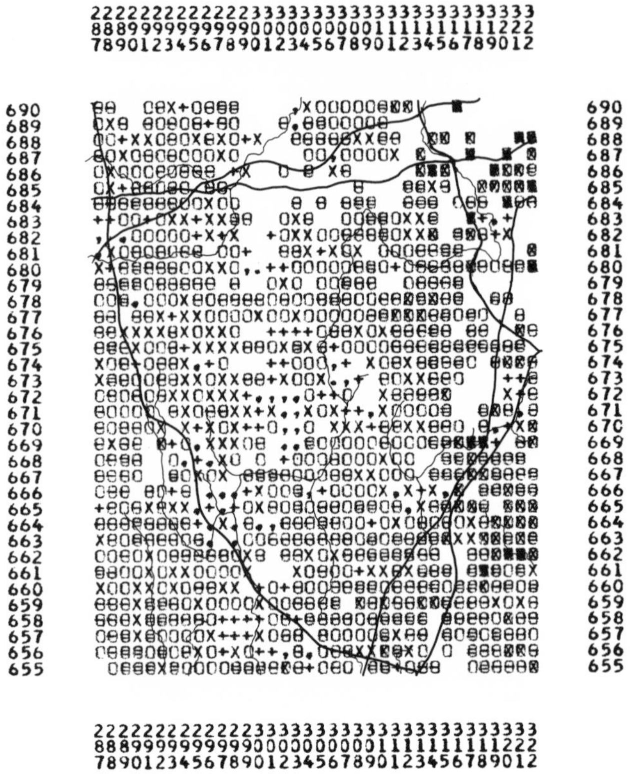

Figure 8
Attractiveness for high-cost housing.
$$y = 3.62 - .82x_1 - 1.77x_2 + 1.88x_3 + .14x_4 + 2.58x_5$$
where

y = attractiveness for high-income housing
3.62 = intercept
x_1 = access to limited access highway
x_2 = access to Route 128
x_3 = distance to elementary school
x_4 = visual quality
x_5 = percentage of area well drained
. 0 = lowest attractiveness
■ 9 = highest attractiveness

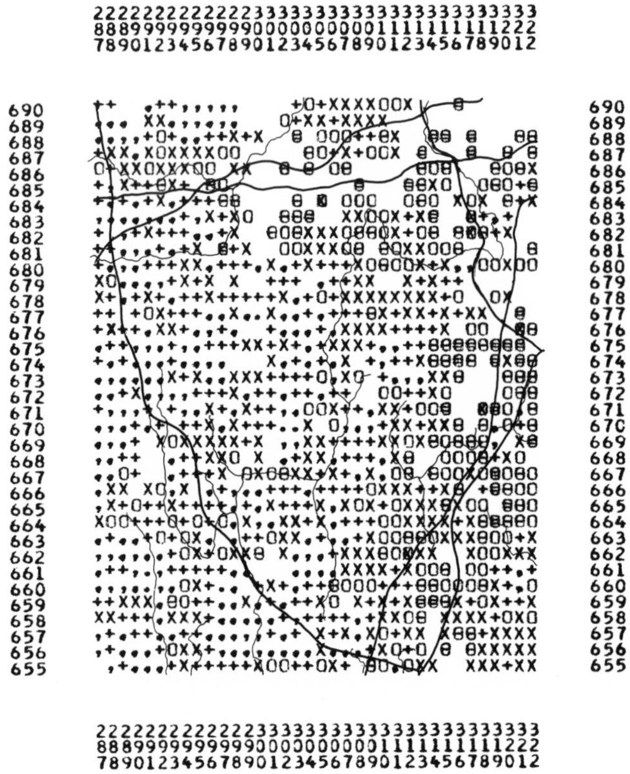

Figure 9
Attractiveness for medium-cost housing.
$$y = -5.67 + .03x_1 - 1.70x_2 - 1.13x_3 + .16x_4 - 1.42x_5 + 2.37x_6$$
where

y = attractiveness for medium-income housing
−5.67 = intercept
x_1 = school expenditure per pupil
x_2 = access to Route 128
x_3 = access to Providence
x_4 = visual quality
x_5 = topographic slope
x_6 = percentage of area well drained
. 0 = lowest attractiveness
■ 9 = highest attractiveness

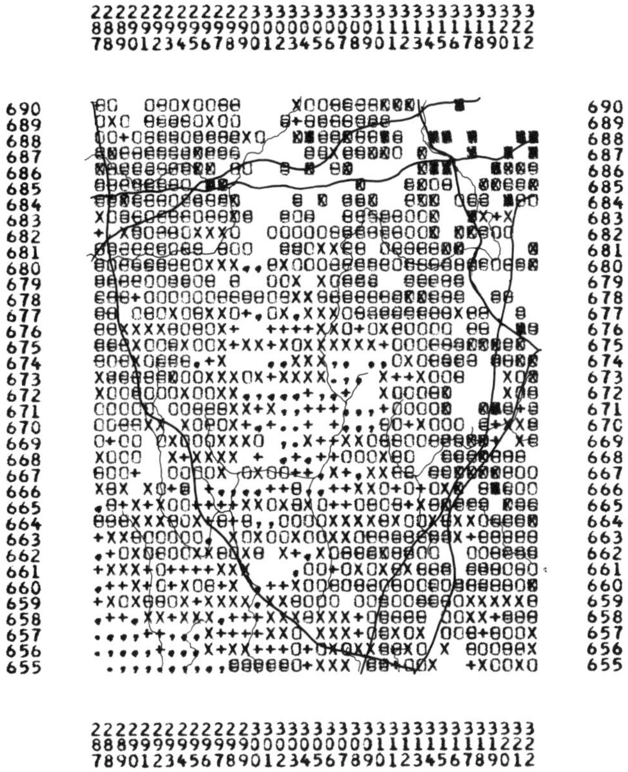

Figure 10

Attractiveness for low-cost housing.

$$y = -4.80 + .03x_1 - .51x_2 + .50x_3 + .007x_4 + .49x_5 + 1.17x_6$$

where

y = attractiveness for low-income housing

−.480 = intercept

x_1 = tax rate

x_2 = access to Route 128

x_3 = access to Framingham

x_4 = school expenditure per pupil

x_5 = topographic slope

x_6 = percentage of area well drained

. 0 = lowest attractiveness

■ 9 = highest attractiveness

Percent of Population
by Income

	Low	Medium	High	
Low	0	50	13	63% at 15 persons/acre
Medium	10	10	5	25% at 40 persons/acre
High	5	5	2	12% at 80 persons/acre
	15%	65%	20%	

Density

Percent of Population
by Income

	Low	Medium	High	
Low	9	7	6	
Medium	8	4	3	1 = first priority
High	5	2	1	9 = ninth priority

Density

Figure 11
Residential demand.

Figure 12
Residential allocation priorities.

density, and that 15 percent would be low-cost, 65 percent middle-cost, and 20 percent high-cost.

In the allocation process, one of our major assumptions was that the housing market is one in which high-cost and high-density development could outbid any other type, then high-density and middle-cost, etc. (See Figure 12.) In each time period we were given a population that would represent our total new demands. We allocated housing types beginning with the highest level of value and exhausting the land supply at each level of value on down. In bidding for residential land, we knew what land was available and which land was most attractive for each type of housing by cost. Having made the allocations, we intended to update the data bank after each simulation iteration. Because of time limitations, however, updating was done only after simulation stages T–3 and T–5.

An important issue, we discovered, was the conflict between producer and consumer attitudes toward residential allocation. This conflict has been discussed by several research groups, such as Chapin's at the University of North Carolina. Our regression models generally described consumer aspects of residential location. We took it upon ourselves, in the role-playing of the allocation, to simulate the producer aspect of residential development. In other words, we were to act out the roles of the people who would actually produce the housing, and their constraints and values would be imposed on the results of the consumer analysis. This role was often in conflict with the training we had had as architects, urban designers, and landscape architects, which had urged us to act in the interest (or what we think is the interest) of the common good, the public good, the landscape good, or the regional good. Now we were being asked to play roles to satisfy a private good. Therefore, in using a profit motive, we established a concept that we called the "value" of a cell. This is a complex idea, and we mulled over it for many weeks, in order to try to define the abstract value that land has beyond its actual financial value—call it its amenity value. When a producer considers developing a site, he weighs its total value against what he can get as a return by produicng a commodity —housing—on that land. He tries to maximize his own profit, the difference between existing value and the value that he can bring to it. This is the role that we were trying to play. For example, in the allocation process, if we found that in a previous iteration time a cell was allocated for conservation, this would make the value of a neighboring cell that we were considering higher because we knew that we would have this breathing space, a major free amenity. So we would jump in and grab that cell before more of it could be committed to conservation. This, of course, would be opposite to the kind of training we're given in school, where one would never encourage this kind of thing. We had a whole series of very heated arguments about cells which had amenities, like lake fronts. As residential producers, we immediately came in and bought all lake fronts because we saw residual profit value in obtaining lake front land as soon as possible in an area that was going to be heavily urbanized. The conflicts are obvious.

Often, interesting things developed at that point. For example, after we had allocated a certain cell, let us say on a lake front, and the public evaluations from the other models—political, fiscal, visual, and ecological—had told us, "We do not want you there," then we as producers said, "Well, we can hire the best design talent available." We changed roles at that point and said, "Now we are the designers for the developer, and we say that by skillfully situating our houses and by deeding a certain easement along the lake front, preserving the trees and so on, by such and such an arrangement we can produce housing that will satisfy the public good also." In other words, "You will be able to go fishing on the lake and not see houses all along the shore—you'll see a tree line." This kind of activity took place frequently, with the teams called upon to play advocates both of the private profit motive and of design in terms of the public good. You will see this more explicitly as we demonstrate the actual run-through of the models when these conflicts are resolved.

We tried to avoid collusion with the other teams before the allocation as much as possible. Again, in terms of our training, it would seem natural for the housing allocator to get together with somebody in conservation and somebody in the political process to work problems out beforehand. It does not work like this in the real world, and we were very careful to avoid this kind of collusion.

There were obvious shortcomings because of time and the limitation of the data bank. Another major shortcoming was that we could not make a continuous updating on land costs. For example, if certain amenities were developed in an area, land costs would rise around it. Other development might depress value. These things could not be taken into account after each iteration. We could not make an allocation that would be up-to-date in terms of land costs; and since the whole allocation was based on this notion of value, it makes the allocation suspect in those iterations following an iteration in which the data bank was not up-dated.

Question. When you decided on the percentages of high-cost housing or low-cost housing, did you not make the assumption that high-income groups would live in high-cost housing?
Answer. That is generally what happens.

We said that 65 percent of the people are going to opt for middle-cost housing. We could have said that 65 percent of our housing will be middle-cost housing, but not necessarily that 65 percent of the population will be middle-income people.

The point raised is a very interesting one, and one that we discussed. If you follow the mechanics, what it means is that the whole prospect of renovation and renewal, the trickling down of housing, will have to be introduced into the model. And it certainly should be, but it means that in subsequent five-year up-dates we would actually have to introduce chips which would say that according to the "trickling-down" theory, a previous five-year period's high-income housing has become middle-income housing. This adjustment would have to be made at each phase and would have been too complex for us this time. However, it is a very valid point.

Recreation and Open-Space Model (Allocation)
David Gates, Timothy Murray, Susan Smith, Thomas Vint
For our model, Figure 13, it was decided to group recreation and open space into three different types of land use. The first is town-oriented, which includes recreation facilities to meet the demands of the population within each town. The second group of recreation facilities was considered regional. Here the demand is from our study area plus the area immediately adjacent to it. The demand population for this category was taken as one quarter of the population of Boston plus the total population within the study area, a range from T1 = 785,000 to T5 = 1,250,000. The third type was called natural and included any conservation land and sites that we would purchase, protect, and leave in their natural state.

We first determined which of the cells were potential recreational and open-space lands. We eliminated those cells which were more than 50 percent residential or had in them certain types of institutions, a high level of industrial or commercial development, certain environmental nuisances, and either a limited-access highway or an interchange. What was left over was assumed to be available for recreation and open-space use.

Then we went through the data bank and selected a sample of existing developments of the three types—natural, regional, and town. In the absence of data on the recreational qualities of existing sites, we had to turn to experts. We took this sample to the Metropolitan Area Planning Council (MAPC) and asked them if they could assign quality ratings on each of these sample sites from both esthetic and user points of view. The sample cells were thus rated from 0 to 9, with nine high and zero low, and this quality rating was taken as our dependent variable.

Then we developed what we called a facility matrix, the aim being to determine from the data bank those variables which were critical in explaining the quality of each of the three types. Of those, the team selected 8 to 10 variables which were assumed to be most significant. We then performed a step-wise multiple regression analysis on these variables for

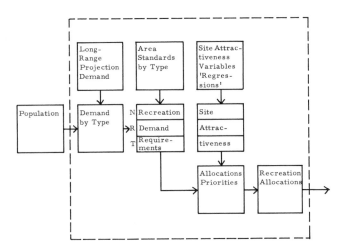

Figure 13
Recreation and open-space model (allocation).

Table 5
Recreation Model: Multiple Regression for User, Regional and Natural Recreation Types.

Independent Variable	User		Regional		Natural§
	Regression Coefficient	% Increase in R^2	Regression Coefficient	% Increase in R^2	Regression Coefficient
Percentage of land well drained	.521 (.228)‡	41.47			
Population of town	.003 (.001)	12.55			
Percentage of town with income less than $3,000	−.372† (.223)	4.77			
Access to major highway			.460† (.299)	3.37	
Time to Route 128			−.463 (.199)	8.22	
Visual texture			.982 (.255)	61.24	
Percentage of poor agriculture					−.44
Percentage of forests					.33
Major water type					.23
Water quality					−.36
Intercept	3.905		2.723		4.34
Multiple determination coefficient (R^2)	.588		.728		

$n = 28, F = 11.413$ $n = 23, F = 16.97$

*Attractiveness measured by use is the dependent variable.
†All coefficients are significant at the .05 level except those marked with an asterisk.
‡Figures in parentheses represent the standard errors of the regression coefficients.
§There was inadequate information available on this analysis at the start of our simulation. The coefficients shown here represent an *ex post facto* analysis of the natural recreation that was allocated in the first three iterations of the first simulation.

each type of open space development. Table 5 lists these variables and summarizes their relative influences.

Regional recreation was regressed on 8 variables. We found that 72 per cent of the variance in the sample was explained by three variables—the access to Route 128, the major type of road, and the visual texture and landscape variation. Landscape variation is the number of different landscape types or personalities existing in the cell. Beyond that 72 per-cent explanation, additional variables made very little differ-ence. The regression analysis gave this equation: the dependent variable, the attractiveness for intermediate recrea-tion, is equal to some constant plus the coefficients times the data values of those three variables. We made regression studies for each of the three types of recreation.

These three equations were applied to all the available cells in the data bank. Figures 14, 15, and 16 are the maps which resulted, showing the attractiveness of each cell for the natural, regional, and town-oriented types of recreation. (When viewing these maps, darker cells are to be read as more attractive than lighter cells.) These evaluations were double checked against aerial photographs and the USGS maps.

Since we were working within an assumed budgetary limita-tion, we developed criteria that would define priorities for selecting the relatively few cells that we could acquire from the larger number of most attractive cells. For our allocation process, we began with the most attractive cells, exhausting each level in turn. We decided to make our allocations on the basis of one ninth of the cells. We also felt that we should allocate in continuous groups of cells if we could, and we developed a strategy for allocating in groups of cells, regard-less of the specific types of open-space use. In general, cells with water took priority over cells with none. Finally, in the later periods in the simulation, we began to link the various open spaces, toward the development of continuous open-space system.

There were a few special criteria for each allocation type. For regional recreation, very flat cells with sparse vegetation were eliminated and marshes were avoided. For natural conservation areas, marsh lands were sought out, and a system of wet-land protection was developed. Isolation from other development was sought, and major roads, industry, commerce, and any type of major residential development was consciously avoided. Town-oriented open space was located in relation to density of residences and generally was required to be within one-half mile walking distance of population concentrations.

Probably the most gratifying aspect of the model was that after working with it for a while and allocating directly from the attractiveness evaluations, it was found to make intuitive

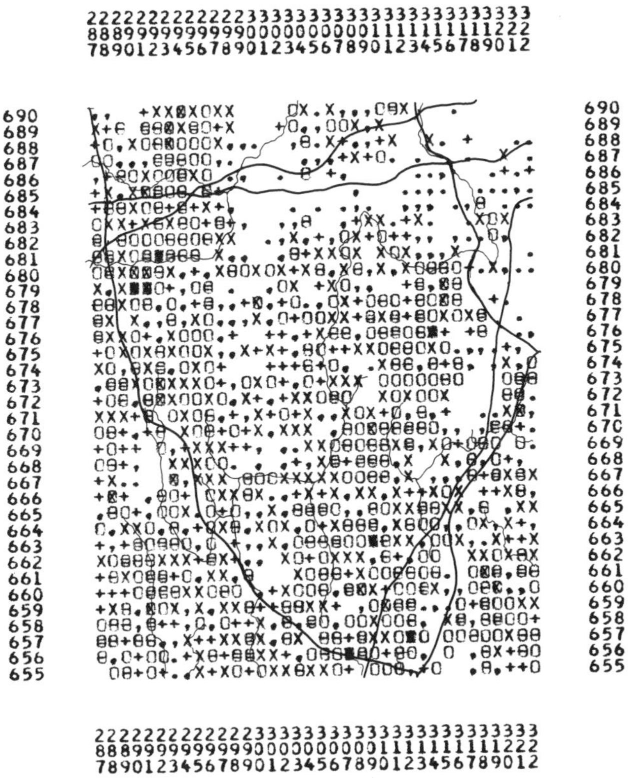

Figure 14
Attractiveness for natural open space.
$$y = 4.34 + .33x_1 + .23x_2 - .44x_3 - .36x_4$$
where
y = attractiveness for natural areas
4.34 = intercept
x_1 = forests, percentage of area
x_2 = major water type
x_3 = poor agriculture, percentage of area
. 0 = lowest attractiveness
■ 9 = highest attractiveness

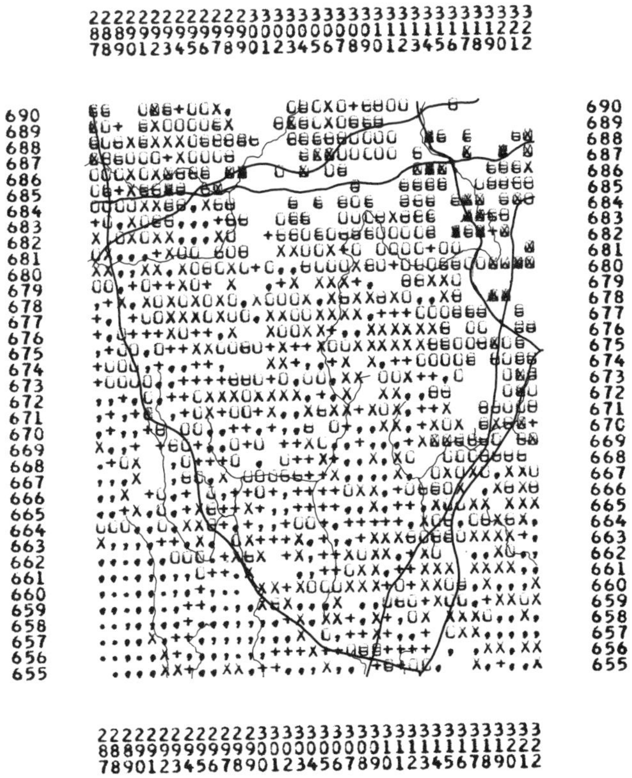

Figure 15
Attractiveness for regional recreation.
$$y = 2.72 + .46x_1 - .46x_2 + .98x_3$$
where
y = attractiveness for regional recreation
2.72 = intercept
x_1 = access to major highway
x_2 = access to Route 128
x_3 = visual texture
. 0 = lowest attractiveness
■ 9 = highest attractiveness

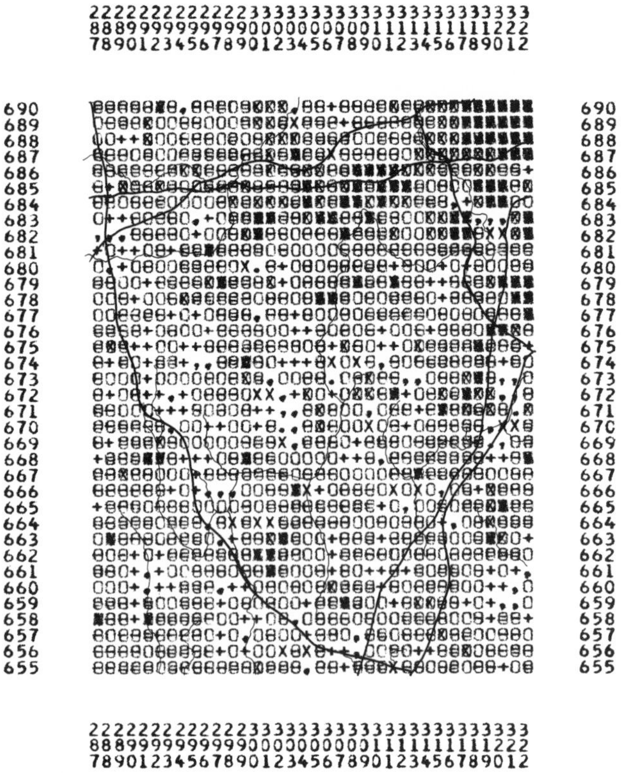

```
222222222222233333333333333333333333
8889999999999900000000001111111111222
7890123456789012345678901234567890 12
```

```
222222222222233333333333333333333333
8889999999999900000000001111111111222
7890123456789012345678901234567890 12
```

Figure 16
Attractiveness for town recreation.
where

$$y = 3.90 + .52x_1 + .003x_2 - .37x_3$$

  y = attractiveness for town recreation
3.90 = intercept
  x_1 = percentage of area well drained
  x_2 = population of town
  x_3 = percentage of town with income less than \$3,000
. 0 = lowest attractiveness
■ 9 = highest attractiveness

as well as numerical sense. Cells evaluated as attractive through our model would also look attractive on the aerial photographs and USGS maps.

On the negative side, we wished we had more time to work in preparing the data for the regression analyses. We had to choose and evaluate a number of sample cells, and the only authority we had for that was, as mentioned, the MAPC, to whom we had to go and say, "How do you evaluate this Lake Cochituate area?" And they would say, "We don't really know how to evaluate it, but basically we think it's good because the people seem to use it." In other words, our evaluations of the samples for the regression are not as rigorous as they could be. We think that could be improved in time, with harder data. Another problem was the one everyone had, that of an incomplete data bank. There were probably more variables that could have been included in the model, and their inclusion could have given us better predictive models. Another factor is that this is a simulation model, which merely projects what exists. The factors that originally decided the locations of existing recreation areas were built into the system. We did not have a chance to question these.

Basically, however, the model worked very well for our sector.
Question. How did you make a satisfactory adjustment for the impact of industrial and residential locations on recreational open space?
Answer. First, cells are no longer available for use. If a previously unused piece of land is spoken for in, say, the second stage, it is no longer possible for open space in the fourth stage. This is one relationship. Another is that to the extent that recreation locations are dependent upon proximity to residence, the new pattern of residence is considered in each allocation stage.
Question. So the chain reaction of the impact of residence or industry on open space and future recreational use is built-in?
Answer. Yes. We might add that it is built-in far more strongly in the evaluation models. The political and fiscal models are updated at every stage. But in the basic allocation models, there was simply not enough time to update at each stage.

Commercial Centers Model (Allocation)
Fritz Stuber

The original intention was to locate commerce and public institutions individually by type. We soon found out this was impossible because of time and the nature of the available data. We therefore had to concentrate on the location of new commercial centers to meet the demands generated by new residential and industrial development. It was assumed that public institutions and service facilities would in turn cluster about these centers in relation to their market area size. (See Table 6.)

Our first assumption was that every town is sufficiently equipped with retail facilities, service facilities, and public institutions for the existing population. Another assumption was that every family in the region has access to a personal vehicle. A third assumption is that all new residential development cells are equipped with the necessary local services and institutions, such as public schools or kindergartens or small-scale community facilities. The allocation of our commercial center is, then, dependent upon new residential development. In some cases, where new residential development and large-scale new industrial development were close together, the employees of the plants were considered potential customers for new retail centers.

The procedures which were used for this model, Figure 17, are as follows: First, new residential demand is created by the allocations of housing and industrial teams. Existing residential demands were assumed to be satisfied. The minimum demand population for the allocation of a commercial center was defined as 4,000 people, generally 4,000 people in new housing in a relatively contiguous district. However, the center type we preferred, Type 3, required a population minimum of 10,000 people. This is the population necessary for a shopping center with the other facilities that typically cluster around them, such as recreation facilities, playgrounds, cinemas, public health centers, restaurants, department stores, and even schools and public administration centers. In some cases—mainly in the earlier stages of the simulation—if the housing team allocated several cells in an area without retail facilities, we considered the speculative allocation of a center with the assumption that the new area would develop further in later stages.

Location criteria for the commercial centers are access to the demand population in terms of time and distance; general access to transportation, to highway interchanges, or to major roads; high visibility from highways; low site costs similar to those of industry (Figure 18); a minimum area for development; and room for future expansion. It was possible in most cases to provide for this last criterion (expansion) because our initial allocation (24 acres or one-ninth cell) was generous. For example: one center was allocated in a town whose area included only four or five thousand people. A second center was developed a few miles away. The region grew very quickly, to over 20,000 people. It was very difficult in the later stages to define which of the two centers would grow further, because the distance between them was so close.

One other thing that should be mentioned here is that there is no special physical form required by the allocation. It could be a shopping center, a strip development, or a cluster of individual stores.

Table 6
Public Facilities by Size of Center

Size 1 300–600 persons minimum	Size 2 4,000 persons minimum
(Size 1 is assumed to be part of any new residential development) day care for baby sitting totlot playground vacant lot workshops laundry shops etc.	recreation public health mental health family service legal and consumer service workshop branch library information center convenience center shopping (drugstore) etc.

Size 3 10,000 persons minimum	Size 4 30,000–200,000+ persons
indoor pool recreation space—playground police security cinema—community theatre indoor recreation health centers art gallery tennis club playfield department store shopping of any kind hotels schools public administration etc.	resource center hospital-center clinic religious center university central library museum any kind of cultural institutions government center (adm.) police, fire—main station convention center zoo stadium recreation and sports facilities olympic pool etc.

Figure 17
Commercial centers model (allocation)

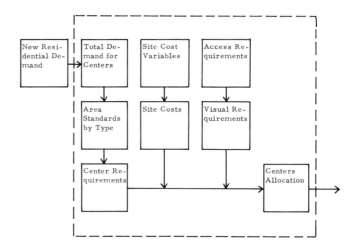

Figure 18
Site costs for commercial centers development.
.   0 = least
,   1 =
;   2 =
+   3 =
X   4 =
0   5 =
θ   6 =
θ   7 =
8   8 =
■   9 = most

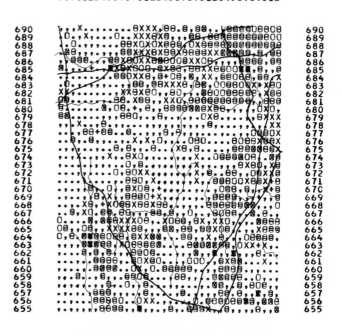

Transportation Model

Nicholas Dines

The transportation allocation process is probably the least glamorous and the least sophisticated of our models. The model represents a very gross attempt to show the general impact of development upon the existing highway transportation network and to estimate the cost to local communities of transportation improvements.

The ideal transportation model is composed of two sub-models: a travel-demand model and a highway capacity model. New development generates new traffic demand. The existing transportation system provides the travel capacity. We assume that traffic demands exogenous to the region are not considered nor is through traffic. We are dealing only with new demands from within the study region. The new total demand volume is compared with existing capacity. If the new volume demand exceeds the present capacity, then the existing transportation must be upgraded. We are further assuming that since the region is already well served with a network of roads, changes will be limited to upgrading existing routes, and we accept as a premise that this existing highway structure will remain for a considerable amount of time. If demand does not exceed the existing capacity, then that volume is subtracted from the existing network capacity. That, in outline, is the ideal model.

To make even such a simple model operational is very difficult, and we did not have time to make the model function automatically. We had to manipulate the structure of the model by hand and develop some rules-of-thumb on which to base our decisions. Furthermore, we wanted to work with a link-node system, while the data bank was organized on the basis of grid cell areas. We therefore used capacity per cell as the basis of the model in use, rather than capacity per road link.

In the model (Figures 19 and 20), as it was developed in use, each cell is described by the highest order road within it. This description, mapped in Figure 20, ranges from 0 (no road) through 9 (the intersection of an interstate highway). This gross rating by type was translated into a volume capacity. Thus, from the number of lanes, we estimated a theoretical design standard and rated the traffic movement capacity for each cell in terms of vehicles per peak hour. This is shown in Table 7. New residential development is assumed to be the major source for travel demand. Each residential density— low, medium, and high—is thought of in terms of numbers of families, with the assumption that each family represents one vehicle trip during the peak hour. Thus, as shown in Table 8, one ninth of a low-density housing cell will add 114 additional vehicles to the existing network during the peak hour and three ninths of a low-density cell will produce 342 vehicle trips. To illustrate the model in use, take a typical cell with a

Table 7
Road Capacity Criteria

Road Type and Capacity	Number of Lanes	Percentage of Use	Capacity in Vehicles per Hour
1	2	50	350
2	2	55	675
3	2	70	600
4	2	75	500
5	3	80	600
6	4	85	1,200
8	6	85	1,800

Table 8
Transport Demand

Housing Density	Number of Families per Cell	Number of People per Cell	Number of People per 1/9 cell	Number of Auto Trips in Peak Hour per 1/9 Cell
Low	1,029	3,600	400	114
Medium	2,743	9,600	1,067	305
High	5,485	19,200	2,133	609

Figure 19
Transportation model.

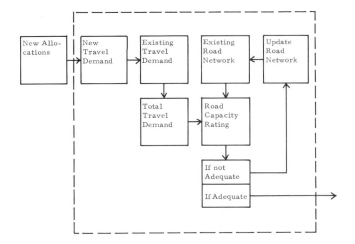

Figure 20
Highway network, T = 0.
. 0 = none
, 1 = unimproved dirt road
ı 2 = paved light duty
+ 3 = paved medium duty
X 4 = urban streets
0 5 = heavy duty
θ 6 = divided with access
X 8 = divided with limited access
▮ 9 = interchange on divided highway with limited access
Source: U.S.G.S. Map, Aerial Photographic Interpretation, and
Mass. D.P.W.

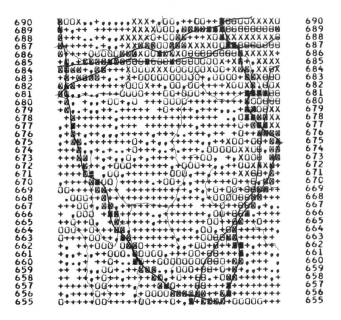

third-level road—which means a medium-duty paved road. The Residential Team allocated two ninths of the cell to medium-density housing. That produced 610 vehicles per peak hour, as opposed to the 600 vehicles per hour capacity of the cell. They thus had exceeded the absorption capacity, and the road had to be upgraded to at least the fourth level, which is the heavy-duty urban street. This information was then given to the fiscal model, and costs were assigned to the towns for the upgrading in terms of capital and maintenance costs.

To summarize, we were unable to effectively handle a links-and-nodes system. We were, however, able to upgrade cell by cell. The model did give an estimate of the road improvements required to meet new demands and the accompanying dollar costs.

Political Model (Evaluation)

David Parry, Ghazi Sultan

This model is an attempt to predict local political attitudes toward new development. The issues evaluated in this political model were those presented by the other models: industry, residence, open space, commercial centers, transportation, fiscal, pollution, and visual quality. The intention is to increase the effectiveness of the city planner in real political conflict situations. The Industrial Team provides an example: they said they could not allocate industry in Dover because the town would not accept it, but they felt that industry would be acceptable to an adjacent town. We wanted to be able to predict the political attitudes of the towns in the study area toward different developing land uses.

People affect new development either by voting or by griping to newspapers, political representatives, or other centers of power. Our model is dealing with griping, because voting is something which happens only every few years. So griping is felt to be the major way that land use can be changed by popular action.

We had to make several assumptions during the development of the model. First of all, we divided the study area in the same way that the fiscal model did—that is, by town—and we assumed that attitudes towards new development of the different income groups within the town would be uniform. Other aspects of the model which required simplification by means of assumptions were the issues related to new development, the characteristics of the districts. Our final assumption was that there are larger districts in the study area where people will tend to have similar attitudes, either because of the geographic location of the district or because of some other consideration, such as an economic one.

Our model includes four assumptions about population characteristics. First were local pressure groups, such as real estate, industry, planning agencies, or government agencies.

These will have real power and will probably affect which land uses go into the area. For instance, woodworking industries on the Charles River would probably have a major say as to whether antipollution devices were or were not installed, regardless of the vote, because they are the people really being affected—and besides, they employ large numbers of workers. The second aspect was income. There is a correlation between personal income and voter behavior. Studies of municipal voting on taxation find that the higher the income, the more people tend to vote for all tax measures, regardless of who benefits. People with low incomes tend to vote against higher taxes unless there are very clear benefits for them (Wilson and Banfield). The third aspect that we considered was the distinction between homeowners and renters. Homeowners pay the property taxes, and most of local development, such as recreation or utility development, is paid out of property taxes. So homeowners tend to vote "no" on development that implies higher taxes. The fourth aspect of population characteristics was ethnic. For example, Negroes, no matter what their income, tend to vote for public expenditures. As Wilson and Banfield say, they have a higher "public regardingness." In the model, as it developed in use, we were only able to use one of these attitude influences. Income was the only factor on which we had good data, and since the residential model allocated housing by income, we could update this variable in the simulations.

The model, Figure 21, as it was developed in use, is divided into four major parts: the population by income, the reaction variable, the power variable, and the issue attitude. Each issue has a different attitude, and each issue is calculated separately for each town at each stage. In outline, we first define the issue to be analyzed, for example, new heavy industry. Then we divide population into the percentages of each income group in the town—percentage high, middle, and low income. Then we find their reaction variable in terms of the intensity of positive or negative feeling toward the development. This is found by a local survey. Last, we add a power variable which we have used as a "fudge factor" in the assumption that this is not a direct popular democracy. All persons are not politically equal. Influence is related to vested interests, and people with higher income probably do have more power in local affairs.

This is how the model was used. First, we chose the issues. In order to develop the reaction variable in the time available, we subdivided the region into what were called established towns, provincial towns, developing towns, and potential-growth towns. In most cases, these were geographically contiguous. The assumption was made that attitudes would be consistent within each of these four groups. We then selected two towns in each category and interviewed two "experts" in

Table 9
An Example of the Political Model in Use: The Low-Income Housing Issue

The low-income housing issue caused the greatest reactions in towns. Except for heavy industry, it was the only development against which people objected, and it was the only development which the political model was successful in rejecting from certain subregions. The reaction variables of each income group towards new low-income housing is shown for each subregion:

	Provincial	Established	Developing	Potential Future
High Income	−2	+2	−3	−2
Middle Income	−2	+2	−3	−2
Low Income	−1	+1	−1	+1

For Southboro, a "potential growth town," there is considerable change in attitude because of a large change in income distribution in the town:

	Percentage Population		Reaction Variable		Power Variable		Political Response
1965							
High Income	20	x	−2	x	3	=	− 120
Middle Income	70	x	−2	x	2	=	− 280
Low Income	10	x	+1	x	1	=	+ 10
			Town response				− 390
1975							
High Income	40	x	−2	x	3	=	− 240
Middle Income	50	x	−2	x	2	=	− 200
Low Income	10	x	+1	x	1	=	+ 10
			Town response			=	− 430

For Wellesley, an "established town," there is little change in attitude over time, because of only a small change in income distribution in the town:

	Percentage Population		Reaction Variable		Power Variable		Political Response
1965							
High Income	60	x	+2	x	3	=	+ 360
Middle Income	40	x	+2	x	2	=	+ 160
Low Income	0	x	+1	x	1	=	0
			Town response			=	+ 520
1975							
High Income	70	x	+2	x	3	=	+ 420
Middle Income	30	x	+2	x	2	=	+ 120
Low Income	0	x	+1	x	1	=	0
			Town response			=	+ 540

each, the town moderator and the newspaper editor. They were asked to estimate the attitudes of the different income groups toward the issues. The reaction variables were determined from the results of these interviews.

The power variable was arbitrarily assigned as 3, 2, and 1 for the high-, middle-, and low-income populations, assuming that high income people would have greater power.

From these factors we calculated the income group reaction to each issue: (high-income population percentage) x (high-income reaction variable) x (high-income power variable) = high-income group reaction. Adding all the group reactions, one obtains the town reaction. When issues were calculated and tabulated for every town, the other teams had a number to work with by which they could tell whether the town was for or against a particular issue.

At each stage, the residential model allocated housing; and we took the new population figures, retabulated, and got new group and town reactions by issue. As shown in Table 9, Southboro is a town in the future potential development area, near the new Interstate Route 495. As the town grows, attitudes change. In 1965, responses were −390 for low-income housing; in 1975 the score for the same issue is −430. This is because there is a much larger high-income group in 1975 than in 1965, with the result that the overall town attitude toward low-income housing becomes more negative.

The only issue toward which the model was consistently useful was that of low-income housing, simply because, apart from heavy industry, this was the only issue on which people reacted strongly. As this model was set up in the overall system diagram, it is an evaluation model and not a generating model, so that it can only say "no" or "yes" to issues such as low-income housing. In several cases, we managed to persuade the Residential Team not to locate low-income housing where it was strongly objected to, and we were able to evaluate alternative locations.

The limitations in the model are those of imprecision due to the broad nature of our assumptions. In our town groupings, we often combined towns which are probably not identical in attitude, such as Brookline and Wellesley. Population groupings by income are also too broad. We should have taken other aspects such as ethnicity and home ownership into account, but we were not able to. Second, it was impossible to get attitude responses that considered quantity of development. For instance, people might not be against the location of a single industry, but they might object if fifty industries were put in the same town. We could not handle this in our questionnaire. Third, our updating was done on population by income, but there was no updating on reaction variables. Since one cannot take a survey of 1975, we had to make very broad assumptions of consistency of attitudes over time within income groups. This is obviously unrealistic although it may be a reasonable operating assumption.

We feel that the model would have been better used as a generating model at the beginning of the development process. It could then have told where to locate industry instead of where not to locate it. This would better parallel local political situations. The towns along Route 495 are asking for industry rather than saying "no."

The main role of the model was as a check; if all groups were in favor of an issue, the model was not used. It became very useful when conflicts developed for land among allocation groups, for instance, if residence and recreation were fighting for the same cell. The model, by showing local attitudes, functioned to resolve such conflicts. (It also had the possibility, which was not fully exercised, of forming coalitions.) In resolving conflicts, someone gains and someone loses, but it need not be completely arbitrary, as was demonstrated by our procedure.

Fiscal Model (Evaluation)

Larry Gibson, David Sinton

The aim of the town finances model was to test the impact of the different types of development on the municipal expenditures and tax rate for each of the towns for each time period. The impact of this development would be measured in terms of changes in the tax rate. The existing trend in tax rates is to increase with new development. Our aim was to prevent large-scale changes in the tax rate, to keep changes within a maximum of one or two percentage points if possible.

We set up a basic equilibrium equation which states that the expenditures equal the tax rate times the tax base plus transfer payments, such as federal and state grants that the town acquired through population changes. From this, then, we can determine for any time period the change in the tax rate. (See Table 10.)

Operationally, we hoped to use this model (Figure 22) to indicate what kind of trade-offs could be made in order to arrive at a fiscally optimum development for a town. For example, if Recreation went into a town, wanted to take two complete cells for recreation, and thereby took money out of the tax base, the result could be a higher tax rate. We might then recommend that they take only one cell or, alternatively, we might suggest a certain residential or industrial input into that town which would add money to the tax base and thus allow them to take out the two complete cells for recreation. As we had hoped, we often became involved in these quantitative trade-offs.

We set up a format, shown in Figures 23 and 24, by which we could update developments after each run, so that it was continually operational at each level.

We worked with three basic assumptions. First, the model

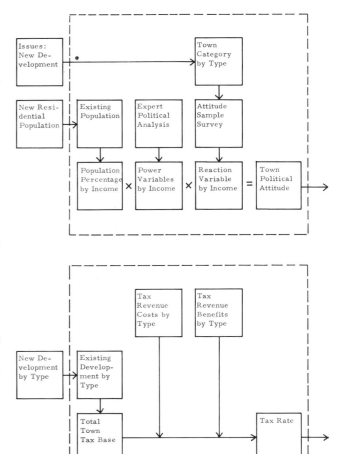

Figure 21
Political model (evaluation).

Figure 22
Fiscal model (evaluation).

TOWN- *HOPKINTON*

	A	T=1 B	T=1 C	T=2 B	T=2 C	T=3 B	T=3 C	T=4 B	T=4 C	T=5 B	T=5 C
A. RESIDENTIAL DEVELOPMENT											
1. Low-Income Units											
a. Low Density	.26										
b. Medium Density	.65	3	1.95			3	1.95				
c. High Density	1.38										
2. Medium-Income Units											
a. Low Density	.24	11	2.64	11	2.64			11	2.64	9	2.16
b. Medium Density	.60	12	7.20	12	7.20						
c. High Density	1.28										
3. High-Income Units											
a. Low Density	.22	6	1.32					6	1.32		
b. Medium Density	.55										
c. High Density	1.17										
B. INDUSTRIAL DEVELOPMENT	0.009	3	0.027					2	0.018	1	.009
C. INSTITUTIONAL/COMMERCIAL	0.0055	1	.005	1	.005						
D. RECREATION/OPEN SPACE		9	—			5	—	40	—		
		1									
TOTAL EXPENDITURE		13.145		9.845		1.95		3.978		2.169	
CHANGE IN TAX BASE IN MILS		+6		+5		+1		+2		+1	

A = CONVERSION FACTOR PER 1/9 CELL (Millions of dollars)

B = NUMBER OF 1/9 CELLS

C = RESULTING EXPENDITURE (Millions of dollars)

Figure 23
Table of expenditures.

TOWN- HOPKINTON	A	T = 1 B	T = 1 C	T = 2 B	T = 2 C	T = 3 B	T = 3 C	T = 4 B	T = 4 C	T = 5 B	T = 5 C	
A. RESIDENTIAL DEVELOPMENT												
1. Low-Income Units												
a. Low Density	3.80											
b. Medium Density	10.1	3	30.3			3	30.3					
c. High Density	22.2											
2. Medium-Income Units												
a. Low Density	7.53	11	82.83	11	82.83			11	82.83	9	67.77	
b. Medium Density	19.0	12	228.0	12	228.0							
c. High Density	40.15											
3. High-Income Units												
a. Low Density	12.65	6	75.90					6	75.90			
b. Medium Density	31.9											
c. High Density	67.6											
B. INDUSTRIAL DEVELOPMENT	36.0	3	108.0					2	72.0	1	36.0	
C. INSTITUTIONAL/COMMERCIAL	1.5	1	1.5	1	1.5							
D. RECREATION/OPEN SPACE	-0.012	9	-0.108			5	-0.60	40	-.480			
TOTAL VALUATION			526.42		312.33		30.24		230.25		103.77	
CHANGE IN TAX BASE IN MILS			-12		-7		-1		-5		-2	

A = CONVERSION FACTOR PER 1/9 CELL (Millions of dollars)
B = NUMBER OF 1/9 CELLS
C = RESULTING VALUATION (Millions of dollars)

Figure 24
Table of valuations.

Table 10

A Model of the Impact of Development on the Tax Rate

Tax Base   The tax base is described by three elements:
M = 1 Real Estate and Personal Property Valuation
     2 Industrial Valuation
     3 Commercial Valuation

Tax base $T = \sum\limits_{M=1}^{3} T(M, I)$ in time period I.

Expenditures   Expenditures are described by seven elements:
N = 1 Administrative
     2 Protection of Persons and Property
     3 Public Works
     4 Education
     5 Recreation or Conservation
     6 Health, Welfare, etc.
     7 Debt Service

Expenditures $E = \sum\limits_{N=1}^{7} E(N, I)$ in time period I.

Development   Development is described by seven elements:
L = 1 Real Estate Development
     2 Residential
     3 Commercial
     4 Institutions
     5 Transportation
     6 Public Service
     7 Recreation or Conservation

Equilibrium Equation $E = (TR) \times (T) + TRANS$,
where
E = total expenditure,
T = total tax base,
TR = tax rate,
TRANS = transfer payments.
From this the tax rate for any time period I may be calculated.

Impact of Development   The tax base will be changed in the period
(I + 1) by Development (L) in the period (I) by the following
increment:
$\Delta T(M, I+1) = A(L, I) \times TAXFAC(L, M)$.
The Expenditures will be changed in the period (I + J) by Development (L) in period (I) by the following increment:
$\Delta E(N, I+J) = A(L, I) \times EXPFAC(L, N, J)$,
where
A(L, I) is the amount of development of type (L) in time period (I),
TAXFAC(L, M) converts development of type (L) to an increment in tax base element of type (M) in the year (I + 1),
EXPFAC(L, N, J) converts development of type (L) to an increment in expenditure type (N) in year (I + J).

tests only changes in the financial state due to new development. Second, the model does not take into account variations which would result from different policies in different towns. For example, Wrentham, which wants industry, might give special tax deals to industry, while Dover, not wanting industry, would not. We could not incorporate this into our model. Third, the model does not take into account variations in transfer payments.

In the actual use of this model, we ran into several problems. First, in developing the specific evaluation factors, we had to work with very general figures that were a number of years old.

For example, how much money does one-ninth cell of industrial development add to the tax rate? How much is the required public expenditure? How much is it for a low-income, low-density housing cell? For these, we had to use the Isard book on municipal costs and revenues, which is about 12 years old. We tried to update these numbers, but in order for the model to work properly we needed a longer period of accurately testing the conversion factors, for saying that one-ninth of industry does generate so much to the tax base, etc. We were never able to adequately verify our figures.

However, there were some aspects of the model that were very successful. We developed a very good updating method, by which we could quickly indicate whether or not the tax base or the tax rate or the expenditures were being raised or lowered too much, and we could give an immediate indication of which amounts of development should be changed. We could get answers quickly enough to be of benefit in the runs while the rest of the simulation process was going on. The operation of the model gave us a method by which we could participate actively in development of trade-offs.

Visual Model (Evaluation)

Peter Jacobs, Douglas Way

The visual model attempted to develop yet another means of evaluation of the development in this region, but it is complicated by the fact that it is based on a nondollar metric evaluation. It is very difficult to put a price on visual quality. Early in the academic year we had developed a visual matrix combining a series of land-use activities and a series of landscapes. These were photographically superimposed in a matrix of all combinations of 15 land-use activities and 11 different landscape contexts. (See Figure 25.) We then did a series of evaluations, interviews, and analyses to determine the visual impact of development on landscape. The model is based on the results of these experiments (Jacobs and Way).

We developed two measures of visual impact on a landscape. The first we called the ability of landscape context to absorb land-use development, in essence, to hide it. This absorptive ability of a landscape was hypothesized as a function of

High ▲

Closed Topography, Heavy Vegetation

Mixed Development, Heavy Vegetation

Mixed Development, Light Vegetation

Rolling Topography Heavy Vegetation

Flat Topography Heavy Vegetation

Closed Topography, Light Vegetation

Closed Topography, No Vegetation

Rolling Topography, Light Vegetation

Flat Topography, Light Vegetation

Rolling Topography, No Vegetation

Flat Topography, No Vegetation

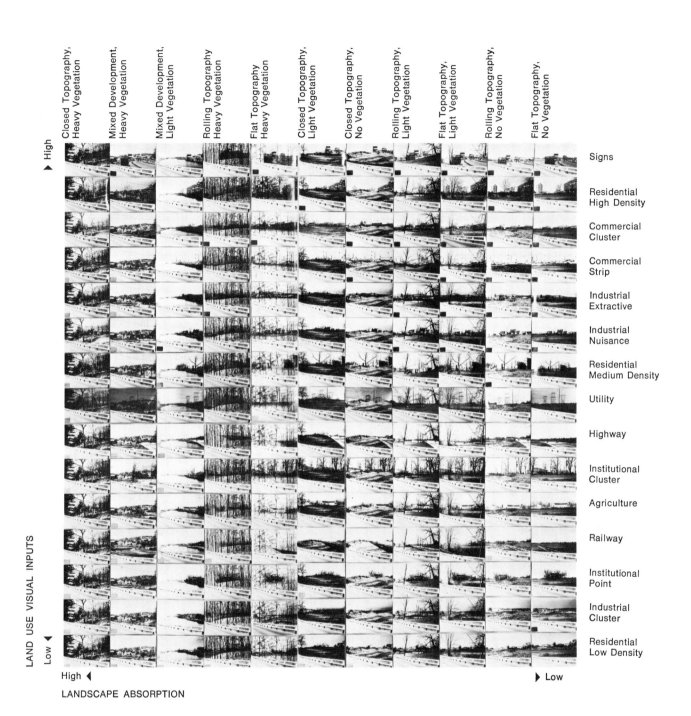

Signs

Residential High Density

Commercial Cluster

Commercial Strip

Industrial Extractive

Industrial Nuisance

Residential Medium Density

Utility

Highway

Institutional Cluster

Agriculture

Railway

Institutional Point

Industrial Cluster

Residential Low Density

LAND USE VISUAL INPUTS

Low ▼

High ◄

Low ►

LANDSCAPE ABSORPTION

Figure 25
A photographic matrix of development in landscapes.

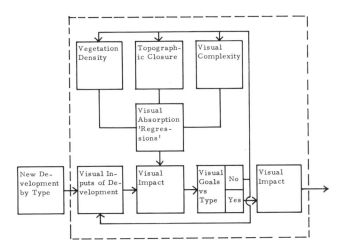

Figure 26
Visual model (evaluation).

vegetation density, topographic closure, and the complexity of the landscape. The second measure, which is perhaps the converse, was the visual impact of each land use upon the landscape. This is determined by the physical characteristics of the activity, be it housing, industry or whatever.

Using these two measures, we developed a visual model for regional evaluation, which had certain assumptions. First, the visual data are assumed to be homogeneous within the cell. We had determined in separate tests that one kilometer is the approximate limit of visibility in a typical landscape. Standing in the center of one grid cell, you are likely to have your view off significantly at the edges. Another assumption, related to the first, was not to consider long-range views. The Prudential Center, for example, has a visual impact that far exceeds one kilometer. But in our study area, such views are less likely to occur.

We worked with a limited number of landscapes and a limited number of land uses, those most common to the study region. There are no deserts, there are no mountains, and it is unlikely that you would find a large-scale mining operation. The model is thus limited to the context of the study area.

The model is independent of value judgments beyond those implied in the goals of the allocation teams. For example, assume that the commercial centers want to be visually conspicuous. The model translates this to say that they want to be in a very exposing landscape. If the Housing Team wishes a rural landscape context which, in essence, masks the housing, then we say that the housing wants to be in a highly absorbing landscape. We find the values through the model and then check them against the goals.

This is the way the model (Figure 26) works through time and as a process. We have three basic categories of data which are analyzed in the initial step: the vegetation density (V), the topographic closure (T), and the visual complexity (C) of each cell in the region. (See Figures 27, 28, 29.) These are combined by the derived formula ($A = 1.46 + .14V + .26T + .40C$) and mapped to show the absorption ability (A) of each cell in the region (Figure 30).

Exogenous to this model, all the teams allocated their basic quantities of land uses to the region. These land-use inputs were recorded by location and then evaluated in our model for their visual impact relative to the goals that they themselves had established. In practice, that is, if a particular cell were highly absorptive and a commercial center were allocated to it, we would evaluate the region immediately adjacent and suggest to the team that they locate instead in that adjacent cell most visually apparent to people passing by. Conversely, if housing were to locate in a cell with no landscape cover or no ability to absorb, we would suggest that they move to a more absorptive cell immediately adjacent or bear the cost of

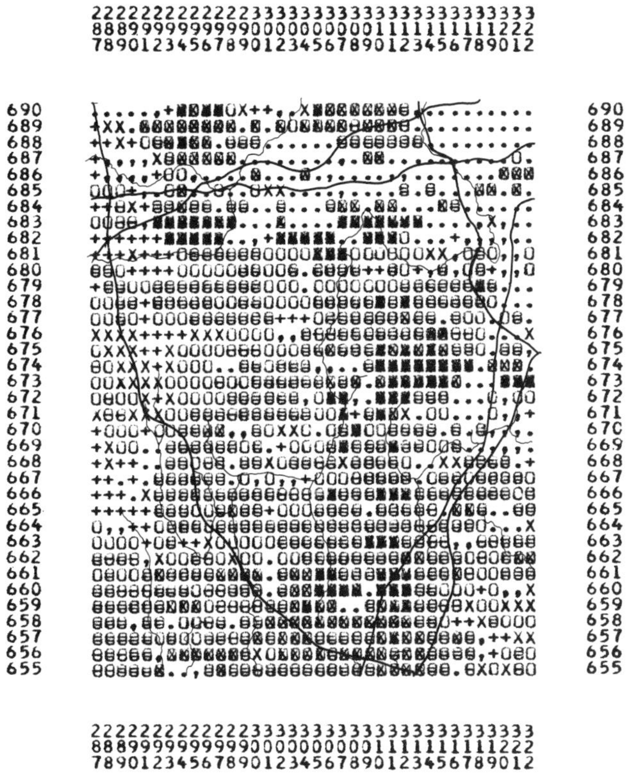

Figure 27
Vegetation density. T = 0.
. 0 = no vegetation
, 1 =
‚ 2 =
+ 3 =
X 4 =
0 5 =
θ 6 =
θ 7 =
▨ 8 =
■ 9 = dense vegetation
Source: Aerial Photographic Interpretation

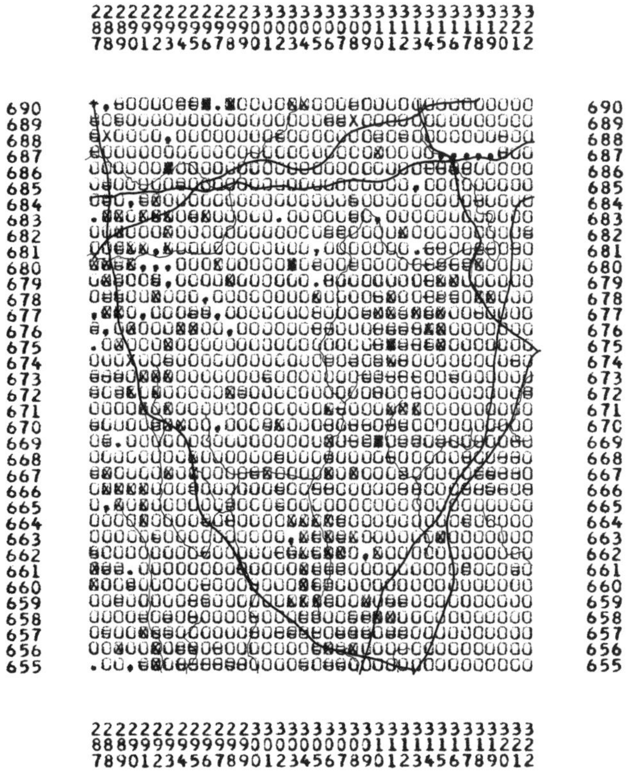

Figure 28
Topographic closure, T = 0.
The measure of a cell by its most absorptive part.
. 0 = most exposed
, 1 =
‚ 2 =
+ 3 =
X 4 =
0 5 =
θ 6 =
θ 7 =
▨ 8 =
■ 9 = most closed
Source: U.S.G.S. Map Interpretation

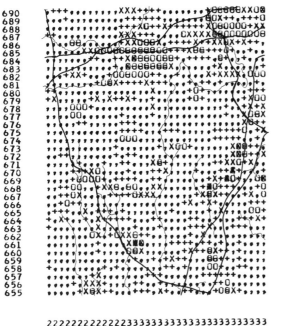

Figure 29
Visual complexity, T = 0.

. 0 = least complex
, 1 =
؛ 2 =
+ 3 =
X 4 =
0 5 =
θ 6 =
ϴ 7 =
⊠ 8 =
■ 9 = most complex
Source: Field Survey, Aerial Photographic Interpretation and U.S.G.S. Map

regrading or planting. If neither the initially chosen cell nor any of the adjacent cells were visually satisfactory to the land use, we might recommend a completely different area. Finally, when all new development has been allocated in this way, we develop a map (Figure 31) which measures the overall visual impact on the region in terms of the predominant visual character. This could be used to test a broad planning goal, such as one that says that the rural aspect of this general region should be protected. We can tell to what degree the inputs of new housing, commerce, and industry have a visual impact on this region, or conversely, to what extent they have been absorbed by the landscapes of the region. Particular areas which are beginning to assert visual change on the region can be seen. This could have an effect upon goals and policies for later allocation of the various land uses.

The limitations of this model are not so much internal but appear in its relation to the processes of development. For instance, there was a very lively argument about lake-side development. A particular cell with a lake was not capable of absorbing development to a large extent. However, the Housing Team wished to locate immediately beside it. It was pointed out that the site was not viable in terms of the housing goal of visual absorption; people using the lake would be constantly aware of the housing. This battle raged on for quite a while until we mutually agreed that the housing group would own the edge of the lake, but with provision of access and a planted buffer zone. We accomplished this by mutual agreement and discussion, but it is questionable whether we would be able to exert any real pressure on a housing developer. In other words, visual evaluation is not yet a tool which has enough substance and power to argue on an everyday reality basis with a housing developer. There is not as yet a legislative or any other mechanism that enables a visual evaluation to trade off effectively with a development goal based on monetary profits.

This explains the process of the model and some of the pitfalls.

Steinitz. I don't quite agree with your political powerlessness. It seems to me that you could mobilize at least conservationists, recreationists, the League of Women Voters, with a prediction of the dire visual consequences of a development.

Answer. We agree that the model has the ability to get out public opinion but inherent in the model is no power whatsoever in terms of development action.

Housing Team. In resolving our particular conflicts, high-cost housing could win the battle but low-cost housing usually lost. We assumed that high-cost housing could perhaps afford the cost of taking an area which was not absorptive enough and planting it and landscaping it in such a way that it would become more absorptive; whereas in the case of low-cost

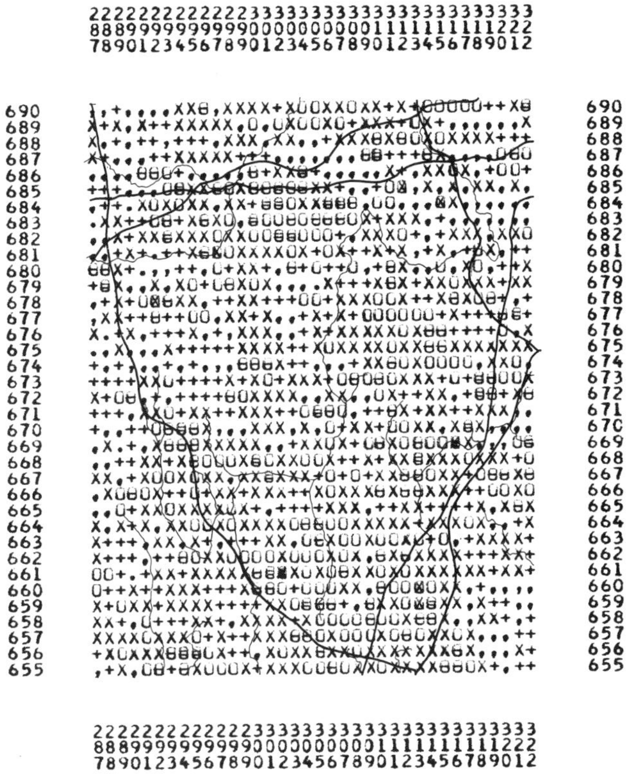

Figure 30
Visual absorption, T = 0.

$$y = 1.46 + .14x_1 + .26x_2 - .40x_3$$

where
y = visual absorption
1.46 = intercept
x_1 = vegetation density
x_2 = topographic closure
x_3 = visual complexity
. 0 = lowest absorption
█ 9 = highest absorption

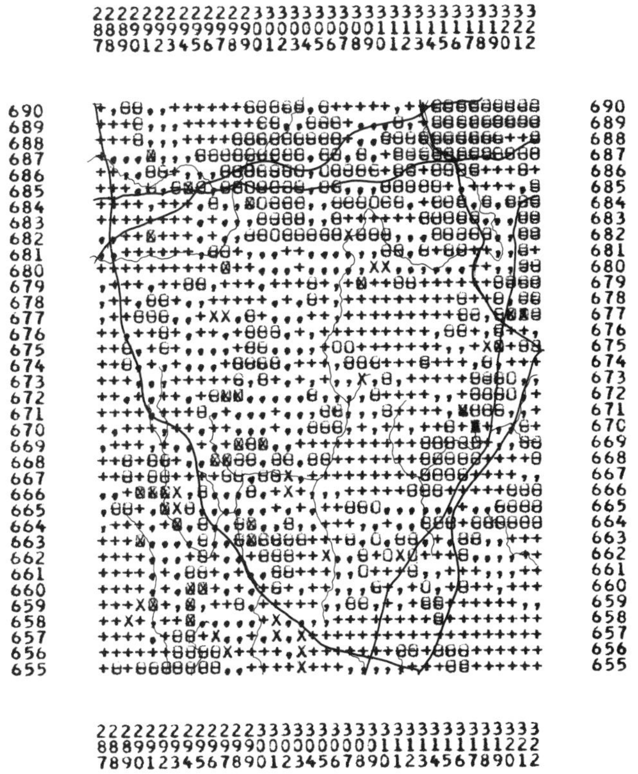

Figure 31
Predominant visual character, T = 0.
, 1 = water
; 2 = wetlands and/or level landscape
+ 3 = forest and/or hilly landscape
X 4 = agriculture, orchards
0 5 = institutions and public services
θ 6 = low-density residential
θ 7 = high-density residential
▨ 8 = commercial
█ 9 = industrial
Source: Aerial Photographic Interpretation

housing, because it's in a situation where every cent is going into the shelter package, we insisted that these units find areas that were initially absorptive.

The Visual Team also won many of the conflict arguments with those of us who were doing the housing allocation, on a very common sense basis. As the process evolved, the visual group ceased being advocates of a vague public good and began to approach us with this kind of argument: "If you follow our advice, you will maximize your profit simply by making better use of the land that you're on, or we can show you an adjacent piece of land which is even more attractive than the one you're on."

Answer. This can be illustrated again by this same lake. The premise is that if you develop right on the lake, the cost of lake lots will increase astronomically, but everything behind it will decrease. Whereas if you save a green buffer strip, so that the lake is protected for everyone, the chances are that the land value of the whole development will be higher on the average.

Question. In this, and in the political model, did anyone study the mechanism of zoning ordinances and try to include these in the process?

Answer. We tried, but we didn't know how to do it. We didn't know how to work out an operation wherein we would propose that all lakes over a certain square footage or in a certain area be protected.

Furthermore, there is a very long time lag in bringing out zoning ordinances on something like this. It would take a minimum of one year to pass an ordinance, and then, under Massachusetts law, any land that has been purchased before the zoning ordinance can be developed for up to the next seven years under the previous zoning ordinance. New zoning ordinances are not very effective in immediate conflict situations.

Steinitz. Future legislation was not part of the simulation process as we developed it, though the effect of such legislation could have been simulated. Also, our data did not include existing zoning maps.

Rogers. The critical thing, I think, in this whole issue is the matter of when is a crisis a crisis. I don't think that you can drum up the political support for an issue of this nature until there is enough evidence of its damage. But how do you determine when the evidence of the damage exists?

Answer. Where one lake in a town that had many was being developed, the chances are that it wasn't critical enough yet. But if all four lakes of one township were being developed, then people might be very concerned. Relative scarcity has something to do with it—we thought about it but didn't quite know how to work it out.

Sasaki. I think you are in danger of making the assumption

that development affects only visual appearance or beauty. It's hard to justify the economic value of beauty, but I think development affects many other things in the landscape, the whole ecological interrelationship, for example, and that too is a real cost. I think your approach is, while excellent, somewhat limited, and I want to make the statement lest we give the impression that development affects only visual appearance.

Pollution Model (Evaluation)
John Gaudette, Larry Lea
In our model, we defined pollution as that impact of man on the ecological system that exceeds the quality standards that man himself has placed on the environment. We will now discuss the general area of pollution, to show how we selected those aspects we used in the model.

We considered four major categories of potential ecological damage: air pollution, noise pollution, solid waste disposal, and water-borne waste disposal. There is presently no problem within our study area with regard to air pollution. There are some air-polluting industries located within the area, but we made the assumption that all new industry in the study area would be put under a strict air pollution code. And we assumed that we need not expect air pollution unless the population density exceeds 5,000 people per square mile.

We did not study noise pollution in relation to the study area because of lack of data.

Solid waste disposal is now treated as an individual or local government problem throughout the major part of the study area. At present, it is not dealt with on a regional basis because there are too many local jurisdictions. Even in the second simulation (metropolitan government), the effects on land use of solid-waste disposal were not large enough to influence the outcomes. So we concentrated on the area we felt presented our major problem, water-borne waste.

Figure 32 shows sewer facilities and their potential capacity for the study area. The black areas are those covered by a treatment system. The white and gray areas are those that are not now covered by any sewerage treatment system and in which waste treatment is mostly an on-site private disposal problem.

The relationship between water-borne waste disposal and the water supply is crucial. Water supply in the region is predominantly from groundwater sources. In Figure 33, the white area relies on groundwater as a water source, and the gray area depends on surface-supply sources. Alternate means of supply from within or without the area are not available at competitive costs. The Charles River, the basin of which is outlined in Figure 33, is limited as a water supply source because of its low median flow. Because of the population increase envisioned by our study, reliance on surface

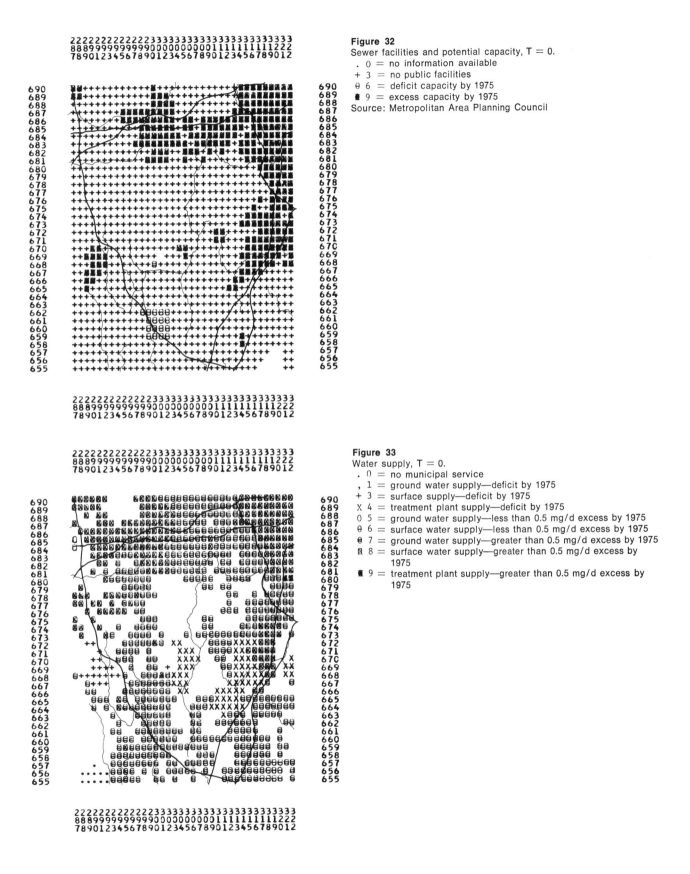

Figure 32
Sewer facilities and potential capacity, T = 0.
. 0 = no information available
+ 3 = no public facilities
θ 6 = deficit capacity by 1975
▮ 9 = excess capacity by 1975
Source: Metropolitan Area Planning Council

Figure 33
Water supply, T = 0.
. 0 = no municipal service
, 1 = ground water supply—deficit by 1975
+ 3 = surface supply—deficit by 1975
X 4 = treatment plant supply—deficit by 1975
0 5 = ground water supply—less than 0.5 mg/d excess by 1975
θ 6 = surface water supply—less than 0.5 mg/d excess by 1975
θ 7 = ground water supply—greater than 0.5 mg/d excess by 1975
▨ 8 = surface water supply—greater than 0.5 mg/d excess by 1975
▮ 9 = treatment plant supply—greater than 0.5 mg/d excess by 1975

Table 11
Per Capita Local Waste Treatment Costs Without Sewerage System

Trickling Filter

Population	Flow (MGD)	Construction ($)	Operating ($)	Annual ($)
500	.0313	135.20	3.02	13.42
1,000	.0625	101.89	2.47	10.31
1,500	.0938	86.36	2.20	8.84
2,000	.1250	76.79	2.02	7.93
2,500	.1563	70.11	1.90	7.29
3,000	.1875	65.09	1.80	6.81
3,500	.2188	61.12	1.72	6.42
4,000	.2500	57.88	1.66	6.11
4,500	.2813	55.16	1.60	5.85
5,000	.3125	52.84	1.56	5.62

Oxidation Ponds

Population	Flow (MGD)	Land (Acres)	Construction ($)	Operating ($)	Annual ($)
500	.0313	1	46.02	1.83	5.37
1,000	.0625	3	38.83	1.61	4.60
1,500	.0938	4	35.16	1.50	4.20
2,000	.1250	5	32.77	1.42	3.94
2,500	.1563	6	31.02	1.37	3.75
3,000	.1875	8	29.67	1.32	3.60
3,500	.2188	9	28.57	1.29	3.49
4,000	.2500	10	27.65	1.26	3.38
4,500	.2813	11	26.86	1.23	3.30
5,000	.3125	12	26.18	1.21	3.22

Source: Richard J. Frankel, "Economic Evaluation of Water Quality: An Engineering Economic Model for Water Quality Measurement," First Annual Report, Sanitary Engineering Research Laboratory, University of California at Berkeley, January 1965.

water would mean building huge storage areas, and the cost of such land development would be prohibitive. All of these considerations have led us to conclude that a model of the effect of pollution on groundwater supply should be given high priority for the purposes of this study.

You are well aware by now that this area is at present basically rural in character, and so the problems being considered in terms of the environment are those of managing the change from rural to semirural to suburban, a transition which makes a tremendous impact on municipal services. The relationship between water supply and water pollution imposes constraints both upon the density of development (which has been discussed by the Residential Team) and upon the technology of disposal. These constraints can be most readily observed in the costs resulting from provision of additional services.

We will discuss waste-water treatment first. A general constraint that we are now facing in this state (and throughout the whole country) is that centralized waste disposal facilities must be provided. There are many federal and state subsidies for waste treatment, but the towns in our study area are small, with maximum populations of only 10,000 people. They are in a very poor position to compete for federal and state subsidies, and this places added stress on the municipal budgets. So the first assumption of our model was that the costs of industrial waste disposal would be borne by industry and that treatment would take place at the specific industrial site. This would therefore not be a municipal cost.

Hence, our water pollution model was most concerned with residential disposal alternatives. We examined three alternatives. The first was individual on-site treatment, in which our bias was toward septic tanks. The second was a system that was not capital intensive. This would be an oxidation pond, which provides minimal treatment without heavy capital expenditure on mechanical and chemical treatment equipment. The third alternative was a capital-intensive system which was highly mechanized and more expensive. This initial analysis is somewhat crude, but in all three alternatives cost is the dominant factor. The residential allocation model brings this out strongly in the additional costs for sewerage collection. In other words, if you have a treatment facility for a high-density, one-ninth cell development, the cost is not dominated by the treatment facility itself but by the sewerage system used to convey the waste from the individual homes to the treatment facility. Tables 11 and 12 show the range of costs for waste treatment per capita for the alternatives of trickling filters and oxidation ponds, in the presence or absence of sewerage systems.

The result of this examination was that we imposed a severe cost constraint on the Residential Team for high-density development in rural areas where there are no treatment facilities (our model furnishes those costs), and therefore the character of the development that took place in the simulations was mainly, in the rural areas, low density. In the case of low-cost housing, the cost per unit was a severe constraint, so that you find that low-cost, high-density development is allocated only to those areas where there are existing sewer facilities. High-income people, on the other hand, could afford the cost no matter where they were, so that high-income, high-density, and medium-density development may be allocated to rural areas. The waste treatment part of the model related to the decision-making process of the residential developer and to the municipal decision-making process.

The model integrates waste-water disposal with water supply. We have in this region a groundwater resource that is limited in its potential for development of water supply. The basic assumption was that groundwater would continue to be the principal low-cost supply alternative. Individual home owners would have pumps, or water would be municipally supplied. Our pollution model, shown in Figure 34, is a physical input-output model. Our input was the rain; our output was the treated waste that was discharged into the rivers or the ground. Septic tanks are a local recharge scheme; people use water and treat it in a septic tank which then leaches it back into the ground where it will eventually be used again. Our decision in using this model was: what pattern or what density of development in a municipality would deplete the groundwater resource such that you would have to import an alternative supply of water? The model assumed a certain infiltration capacity of the soil, which was a function of the density of development in the area.

As it turned out, the water supply aspects of the model had little effect on the simulations because development was generally scattered and of low density throughout the study area. If development had been concentrated in one particular place which had higher densities, a large demand for water, and a lot of area covered by roads and buildings, then you might have had a local depletion of the groundwater resource. The water supply part of the model, therefore, did not impose constraints throughout the whole region for the time periods we considered. This would not, however, be expected to continue beyond those time periods.

Constraints on water quality, as mentioned, are imposed on all stream systems in the United States. The Charles River system dominates our study area; therefore our model focuses on this river basin. You may consider a river as a living biological system and characterize it just as you would characterize a neighborhood. Bacteria interact in a river just as individuals interact in a neighborhood. The Charles River, by our neighborhood analogy, borders on being a slum. It is typified by a very sluggish, low flow; it has many dams; there

Table 12
Per Capita Local Waste Treatment Costs Including Sewerage System

Trickling Filter

Population	Flow (MGD)	Total Costs High Density ($)	Total Costs Medium Density ($)	Annual Costs High Density ($)	Annual Costs Medium Density ($)
500	.0313	170.	181.07	16.11	16.95
1,000	.0625	125.	135.77	12.07	12.92
1,500	.0938	105.	116.23	10.30	11.14
2,000	.1250	94.	104.67	9.23	10.08
2,500	.1563	86.	96.79	8.50	9.34
3,000	.1875	80.	90.96	7.96	8.80
3,500	.2188	75.	86.42	7.53	8.37
4,000	.2500	72.	82.75	7.18	8.02
4,500	.2813	69.	79.70	6.89	7.73
5,000	.3125	66.	77.12	6.65	7.49

Oxidation Ponds

Population	Flow (MGD)	Total Costs High Density ($)	Total Costs Medium Density ($)	Annual Costs High Density ($)	Annual Costs Medium Density ($)
500	.0313	81.	91.90	8.06	8.90
1,000	.0625	62.	72.71	6.36	7.20
1,500	.0938	54.	65.04	5.66	6.50
2,000	.1250	50.	60.64	5.24	6.09
2,500	.1563	47.	57.70	4.96	5.80
3,000	.1875	45.	55.54	4.75	5.60
3,500	.2188	43.	53.87	4.59	5.43
4,000	.2500	42.	52.53	4.46	5.30
4,500	.2813	40.	51.41	4.35	5.19
5,000	.3125	40.	50.45	4.25	5.09

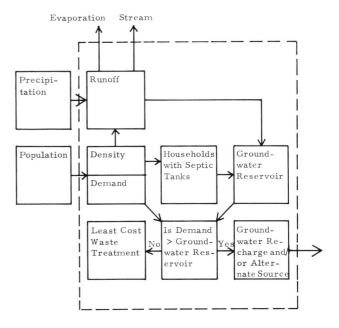

Figure 34
Pollution model (evaluation).

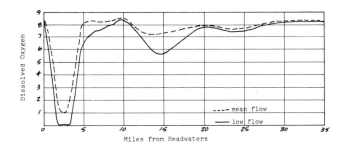

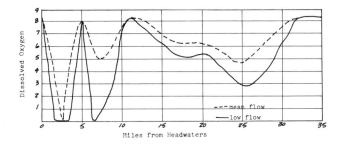

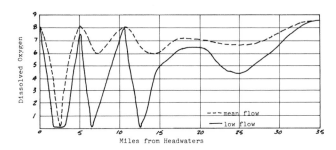

Figure 35
Dissolved oxygen, existing conditions.

Figure 36
Dissolved oxygen, simulation one.

Figure 37
Dissolved oxygen, simulation two.

is a little industrial pollution; and there are large quantities of domestic waste flow.

The one parameter most often chosen to classify streams by federal standards is dissolved oxygen, because dissolved oxygen indicates the degree of bacterial activity within the river. This is the standard that is used in Massachusetts, and it is quite valid for a river such as the Charles. Our model divided the river into reaches, into each of which a certain municipality discharges waste into the river. If the bacteria depletes the oxygen beyond a certain point in any reach, the river can violate the state standard. This is the basis of the constraint that we placed upon development. The model shows the effects of new development on both water resource use and on the required waste treatment facilities.

Figures 35, 36, and 37 show levels of dissolved oxygen, existing and after the two simulations. The lowest level is zero —no oxygen. In this state, the river is smelly and visually unappealing. (For example, the discharge from Milford, a large, growing community with a sewerage system, is sufficient that the bacteria use up the oxygen in the river, and it becomes anaerobic.) Concentration of wastes and rate of flow are interacting functions in river pollution. Spring flows and summer flows are quite different. Usually, design standards are set between the low and average river flows. Drought flow is too severe a condition, and average flow too lenient. State standards are usually set at about 5 milligrams of dissolved oxygen per liter of water—between 5 and 3 for low class rivers such as the Charles. In Figures 35, 36, and 37, we see that we already have a problem upstream. Because of our bias toward septic tanks, river pollution is not as great as it might have been. The development pattern as it took place in Simulation One—even biased in favor of septic tanks—caused violation of the standards, even far upstream (Figure 36). Downstream, the violation is also severe, because there was more high-density development in that part of the region, that is, more sewers and more discharge into the river.

We developed a pollution model which interrelated water supply, waste-water treatment, and river pollution. Our goals were: to point out potential problem areas in regard to water supply; to provide cost figures for waste treatment to the Residential Team and to show where they, perhaps, should locate their residential development in order to take advantage of existing sewerage systems; and to evaluate waste loads that were developed in the area during the simulations to show how they affected the pollution of the Charles River.

Two Simulations of
Metropolitan Growth

The Simulation System

To review, Figure 38 is a diagram of the simulation system, showing how the sector models were linked together. The system begins with a new population to be accommodated. The population increment for each five-year period increases slowly. The figures, shown in Table 13, come from Metropolitan Area Planning Council population projections and are accepted as given. Derived from the number of new people is the number of new jobs.

The first model activated was the industrial location model on the premise that industry is assumed in this system to be a prime mover, with the power to outbid all other uses for land. The number of jobs required was combined with area standards to determine the industrial area required. A separate analysis was made of the labor supply and markets. Subarea demands were determined, site costs were evaluated, and allocation was made of the required industrial sites. That allocation became a factor in the many variables to be considered for residential location.

For the residential model, the same population input was divided into income groups, which made demands for the different types of housing. Analyses were made of the factors for attractiveness, and these values were applied and mapped for high, medium, and low income. The land available for residential development was found, and, through a system of priorities, housing was allocated in the nine density and income categories. The allocation of residence became a factor in the demand for commercial centers.

The new population is also an input in the demand for recreation and open space. This team was the only one that had to make long-range projections because of the need to conserve natural land. They established a twenty-five-year long-range program of acquisition. Again, the team established area standards by type and analyzed them according to town, regional, and natural categories. The supply factors were analyzed, and attractiveness was measured and mapped for these three categories with the land not available being removed. Open-space and recreation land was then allocated.

The commercial centers received as their main input the allocation from the housing model, and this, when combined with new industrial locations and existing unsatisfied residential demand, gave us the total demand of people without adequate service. Standards by population, distance, and type were established, site costs were measured and mapped, and allocation was made of the new and expanded commercial centers.

The demand caused by new residential development was the principal input to the transportation model, in terms of new traffic demands. Existing demands were known, and the capacity of the existing road network was measured against

Table 13
The External Inputs to the Simulations

Stage		Years		Additional Population		Open-Space Bond Issue
T_1	:	1965–70	:	75,000	:	$2 million
T_2	:	1970–75	:	85,000	:	3 ″
T_3	:	1975–80	:	95,000	:	2 ″
T_4	:	1980–85	:	110,000	:	2 ″
T_5	:	1985–90	:	125,000	:	2 ″

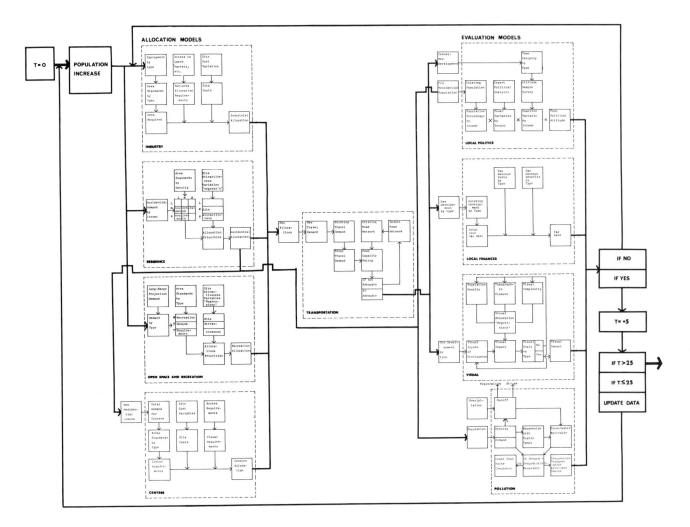

Figure 38
Simulation system.

the new total demands for its positive and negative capacity. If the capacity was negative, if there was an excess of traffic demand over capacity, the existing network was updated and improved. If there was excess capacity, if there was no crowding, no change was required.

These new land uses became the issues in the political model, and the population allocated by the residential model was used to update the population breakdown by income of the towns. The political model was used to measure the degree of political acceptance or rejection of each proposed allocation.

Similarly, the new development became the input for the fiscal model. The existing tax base was known, and tax rates by town were calculated. The tax costs of any new development and the over-all tax-rate change were evaluated.

New development was also the input to the visual model. The existing landscape absorption was known, and the new impact was measured for each type of development relative to its own goals. The change in the visual character of the area as a whole was also determined.

The residential and industrial allocations created a new demand for water which, when combined with existing demands and supplies, could or could not be satisfied. Development caused changes in the runoff coefficients of the land which contributed to flooding. Waste demands were calculated in terms of required treatment facilities and added stream pollution, the goal being less pollution and less flooding.

If the evaluations were satisfactory, i.e., if there was no major political opposition and no big tax rise, if visual goals were met, and if there was no excessive pollution, the data were updated and we went to the next iteration. If any evaluation was unsatisfactory, we had two procedures. First, we corrected the sector itself; for example, if there was a visual objection to residence, residence had the opportunity to make a shift to alleviate that objection. When that was accomplished, it was still possible that there could be conflicts for a specific piece of land among the sectors. These were resolved by the evaluation models. Second, when all sector conflicts were eliminated, the development pattern as a whole was evaluated. For example, if in the process of eliminating an objection for visual reasons, we aggravated the pollution model, development as a whole had to be screened. Only then did we declare that period over, update, and begin again in the next stage. That, in outline, was the procedure as it was developed by the group.

Simulation One
The first simulation was based upon a projection of current attitudes and roles. Figure 39 shows the existing state of de-

velopment in the region. Figures 40 through 44 show the five stages of this simulation.

Simulation Two
With all growth inputs remaining quantitatively constant, the second simulation was run under the following assumption: A Metropolitan Regional Government centered in Boston has incorporated all of the study area. Following from this assumption are these sector implications:
1. *Industrial* development will be concentrated in first priority areas, with a minimum development size of one-half square kilometer.
2. *Residential* development will tend toward larger increments in priority areas. Location of low-income residence will not be constrained on a town basis.
3. *Recreational* locations will not be constrained on a town basis. A policy to develop more continuous open spaces can be considered.
4. *Commercial* concentration can be considered, consistent with residential development.
5. *Transportation* will be updated for the first ten years. Thereafter, consideration can be given to mass transit extensions and/or major road changes.
6. *Political* constraints on a town basis are relaxed as areas of development type are located by least resistance.
7. *Fiscal* evaluation is assumed by the new government level, with appropriate transfers.
8. *Pollution* controls and the provision of water and sewers are assumed by the new government level. Economies of scale and contiguous development will be identified.

Figures 45 through 50 show existing development and the five stages in the second simulation.

Comparison of the Simulations
Steinitz. Would each team briefly discuss how the results were different in the two simulations?
Industrial Team. The allocation procedure differed basically in the two cases. In the first simulation, we assumed that each individual town would be competing for the industry and that therefore each would be able to provide only small concessions. They might be able to build an industrial park, but it would be a small industrial park. We therefore concluded that there would tend to be small amounts of industrial allocation in each town. So in the first simulation we determined the areas of greatest relative attractiveness. And then we used a random process that would determine which zones were going to be built on.

In the second simulation, we made the assumption that metropolitan government, in wishing to control the development of the area, would help develop a limited number of large industrial parks which would be attractive to industry.

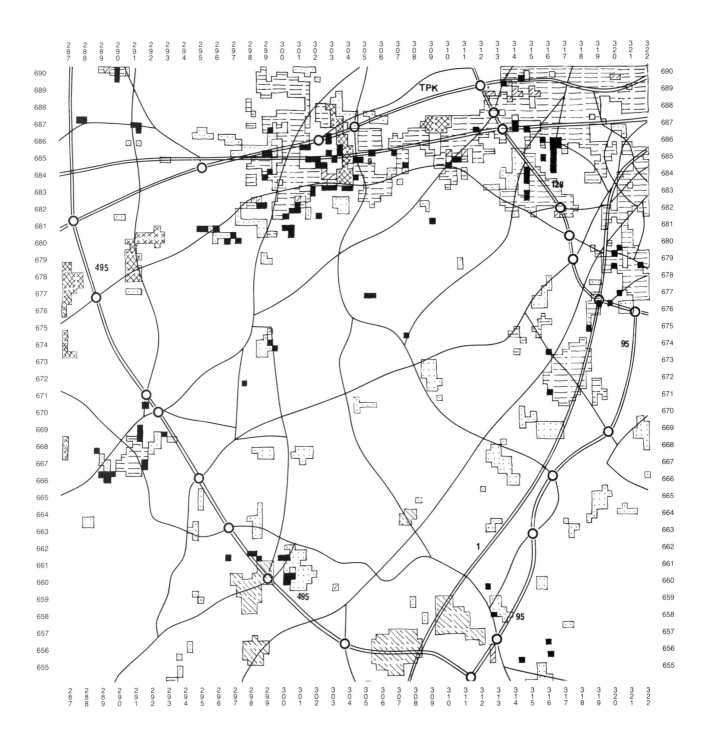

Figure 39.
Existing development, T = 0.

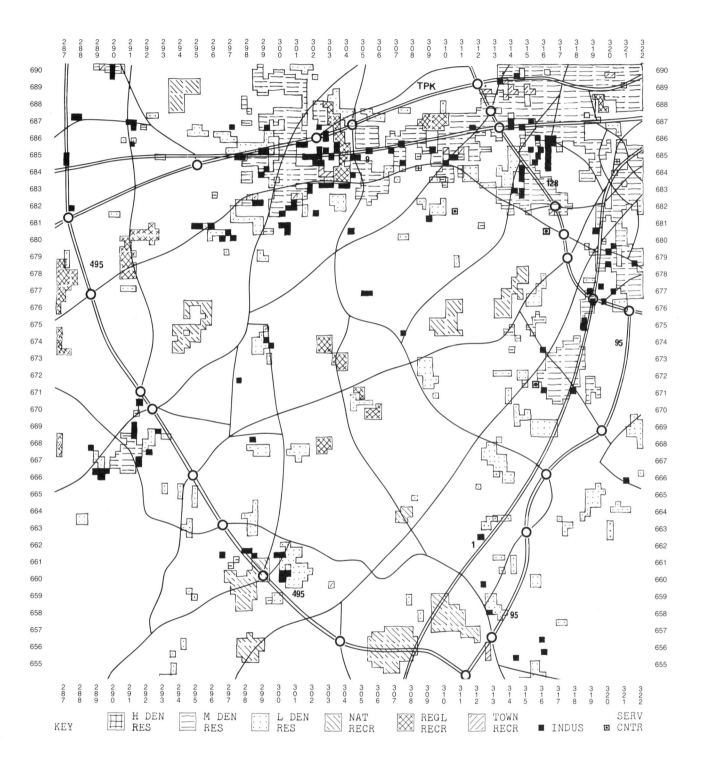

KEY ⊞ H DEN RES ▤ M DEN RES ⠿ L DEN RES ◩ NAT RECR ⊠ REGL RECR ◪ TOWN RECR ■ INDUS ⊡ SERV CNTR

Figure 40
Simulation one, T = 0 + 1.

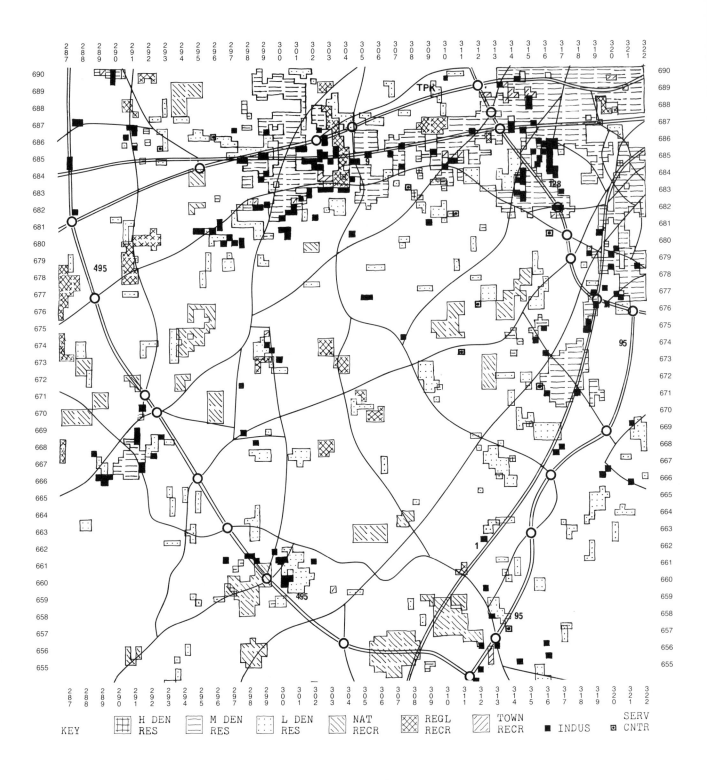

Figure 41
Simulation one, T = 0 + 1 + 2.

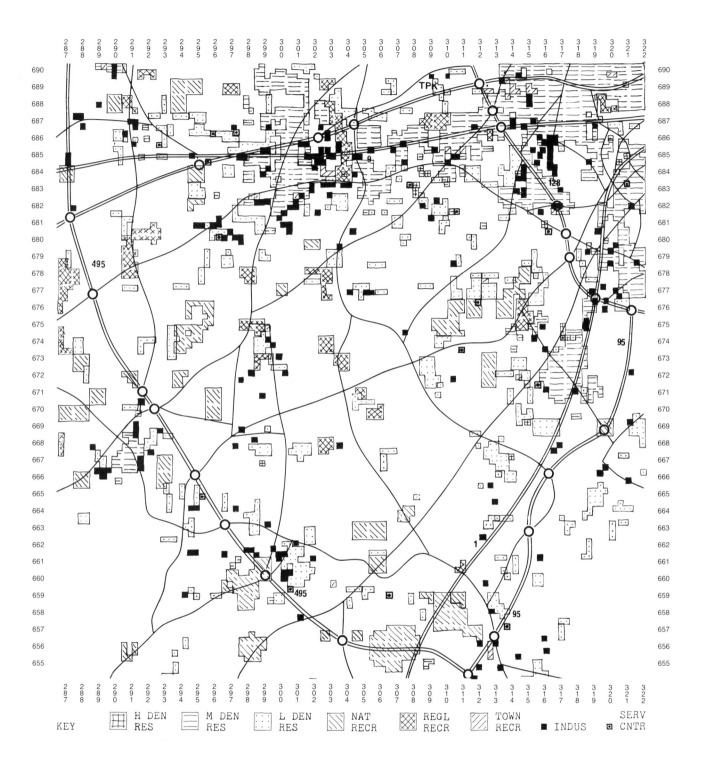

Figure 42
Simulation one, T = 0 + 1 + 2 + 3.

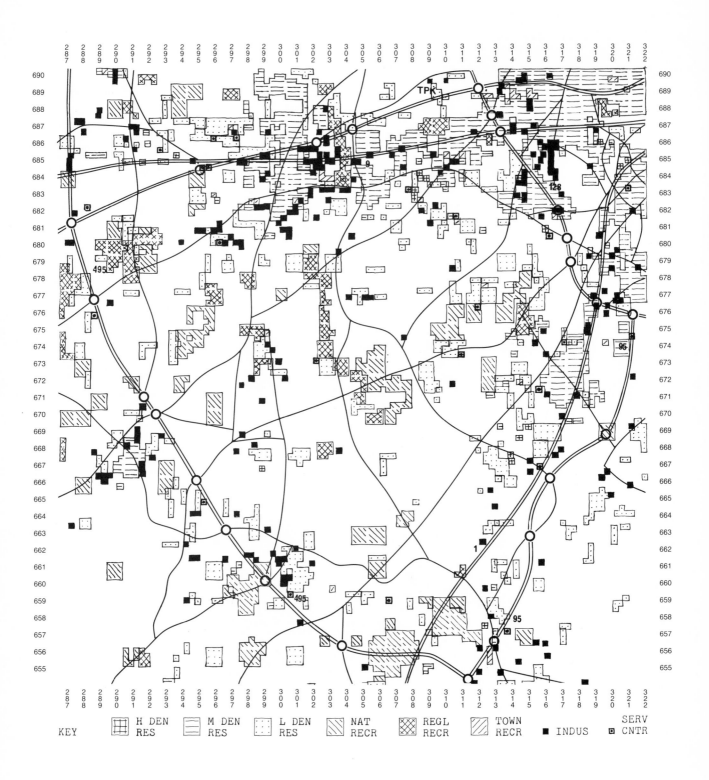

Figure 43
Simulation one, T = 0 + 1 + 2 + 3 + 4.

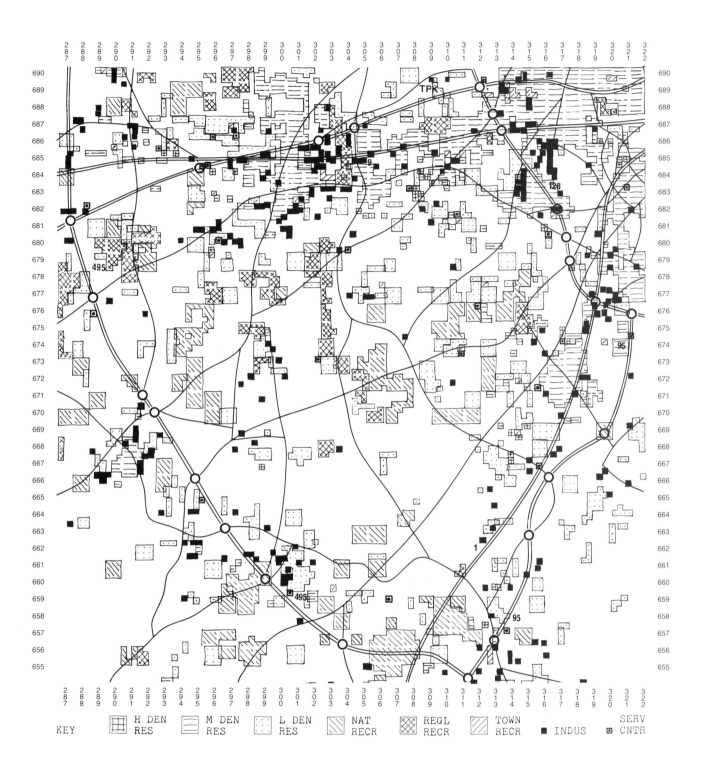

Figure 44
Simulation one, T = 0 + 1+ 2 + 3 + 4 + 5.

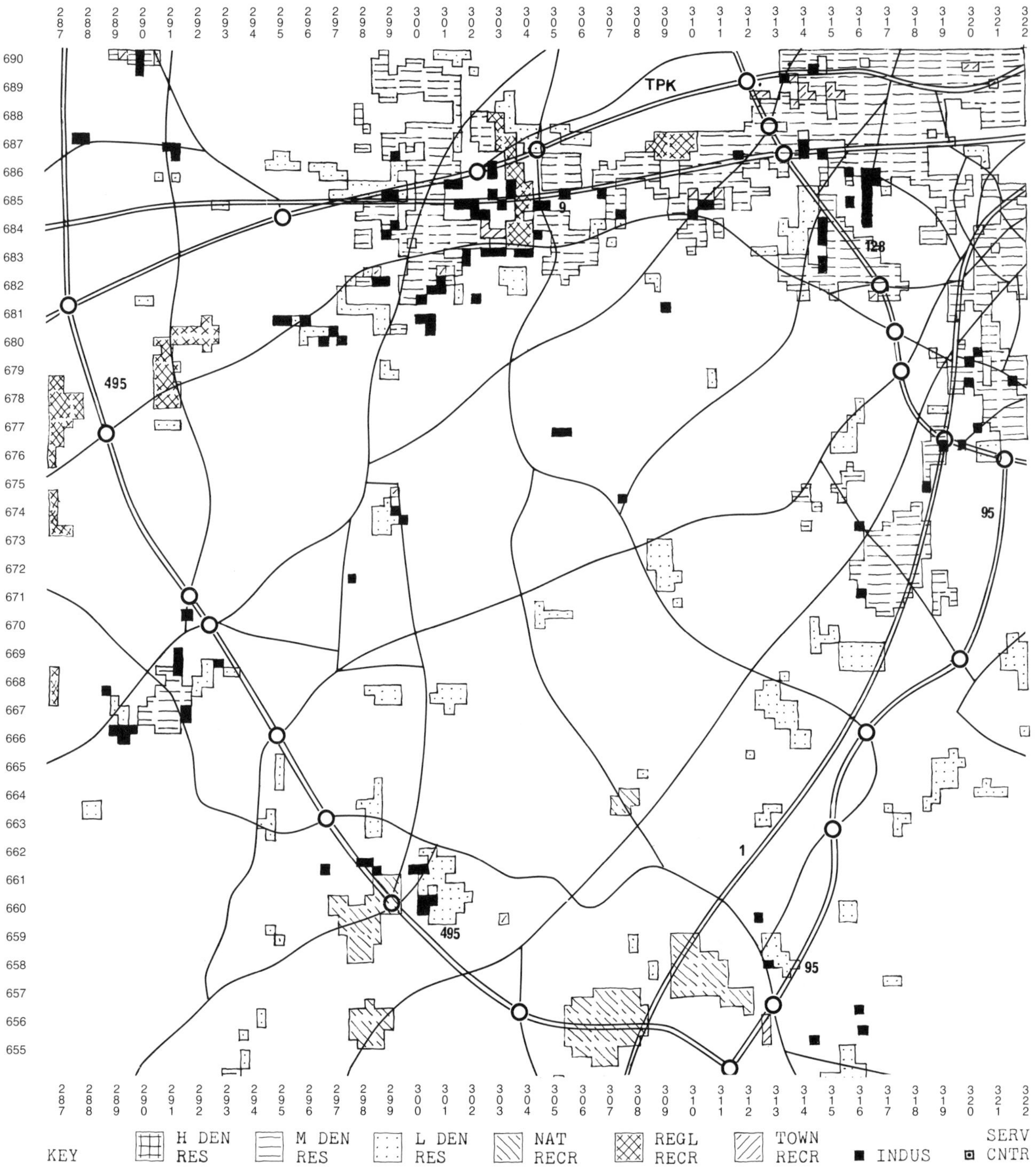

Figure 45
Existing development, T = 0.

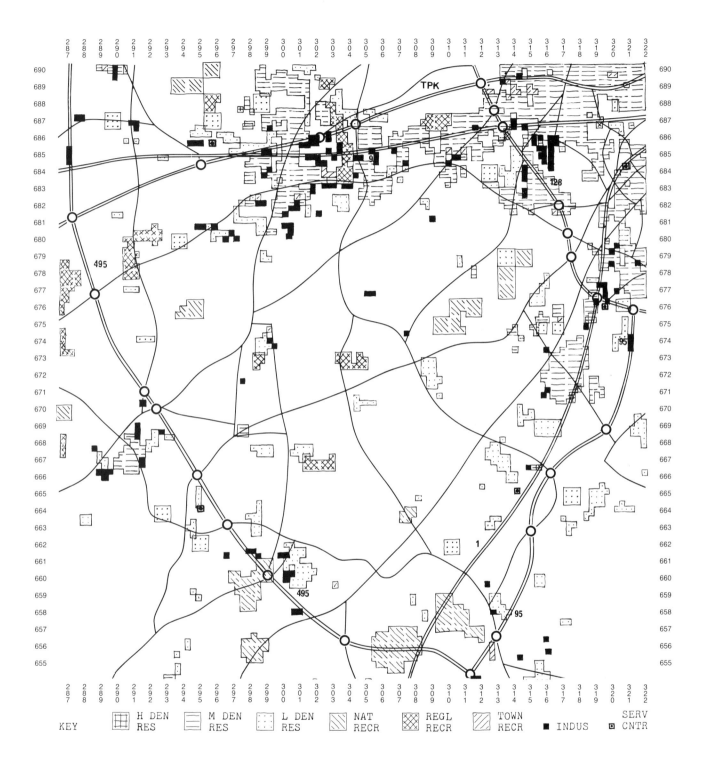

Figure 46
Simulation two, T = 0 + 1.

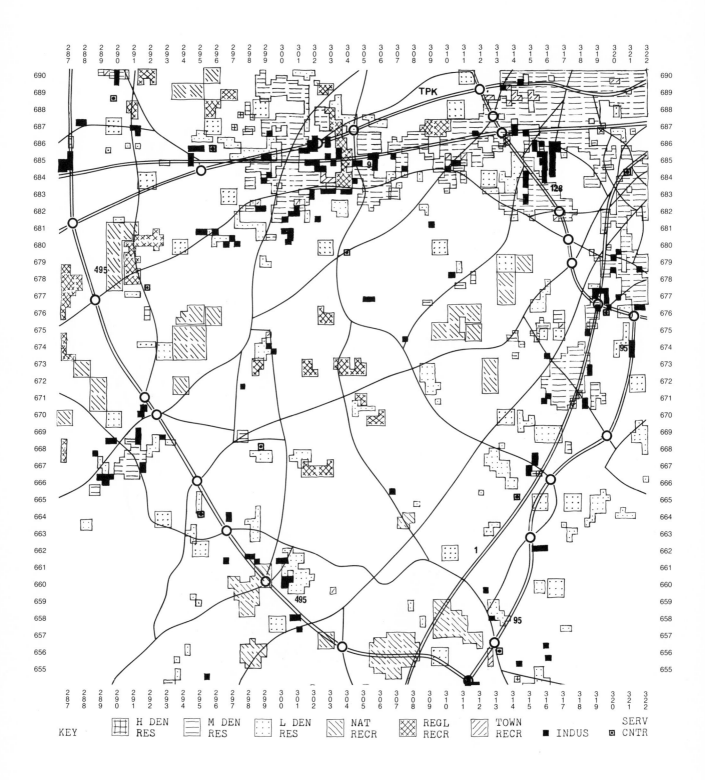

Figure 47
Simulation two, T = 0 + 1 + 2.

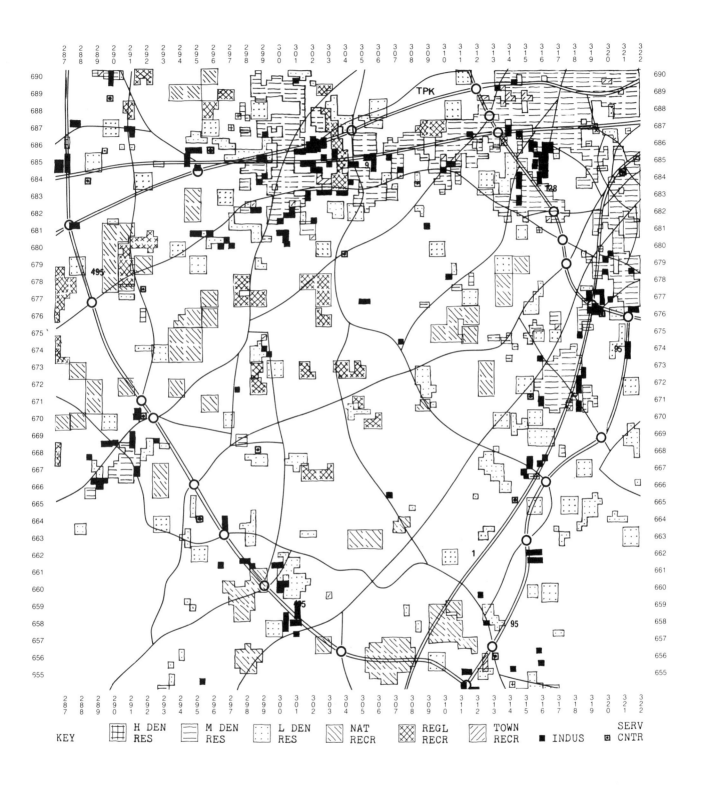

Figure 48
Simulation two, T = 0 + 1 + 2 + 3.

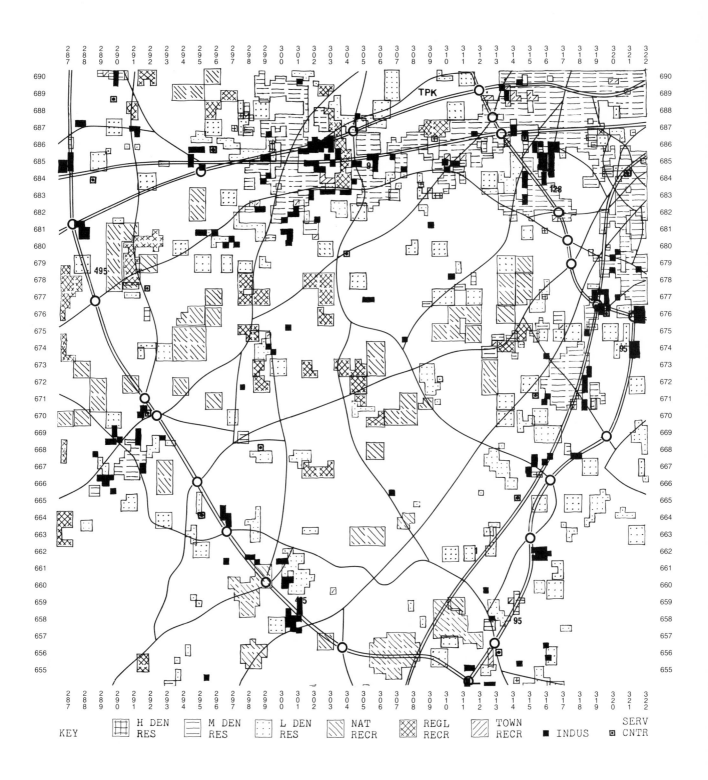

Figure 49
Simulation two, T = 0 + 1 + 2 + 3 + 4.

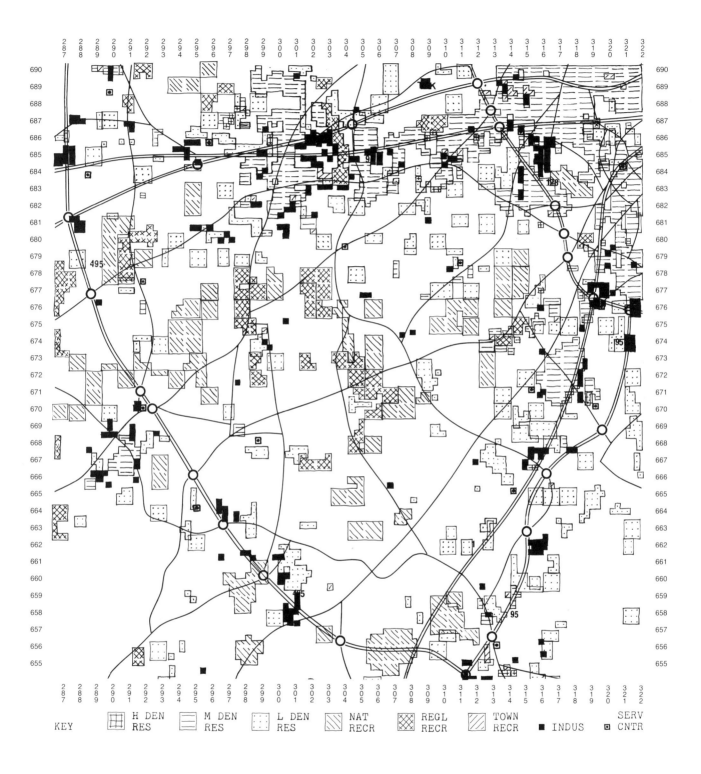

Figure 50
Simulation two, T = 0 + 1 + 2 + 3 + 4 + 5.

Thus, industry would be more concentrated; every time we established an industrial area, it would attract more and more industry over the time periods. Therefore the first time we allocated, we spent a good deal of time discussing where the locations of industrial parks would be. We studied industrial parks in the Boston area, and found that the major criterion for location seems to be their proximity to interstate highways, or limited-access highways. And they are close to interchanges on these highways. We determined where the interchanges were that had suitable land for industrial parks. Once we had determined the location of the industrial parks, we clustered the industry within it. So, in the second simulation, the pattern of industry is much more clustered and very much determined by the transportation network. Just making that one assumption of metropolitan government, which implies a certain amount of planning, you can see how we began to concentrate. Without this assumption, there would have been a much wider scattering of industry and only limited control of where the industry actually situated.

Residential Team. The biggest comparative factor for us was that in the second simulation there was no constraint on low-income housing by town. The other was that we had a larger scale of development, developing more by whole cells than by ninths.

One of the things that was quite interesting to many of us was that the lower-density scattered development of the first simulation had less impact on water pollution than we had thought. When we argued for more concentrated development with a lot more open space, we found that such concentration caused more water pollution than when industry was scattered with appropriate individual treatment facilities.[9]

Recreation Team. The two simulation results were quite similar. In the first simulation, natural open space was often constrained because of our fear of taking too much land off the town tax base. We were able to follow our maps precisely in the second simulation. There were some differences between simulations—as in the town recreation lands—because of differences in the residential allocations. In the second simulation, we found that the town recreation areas were required at a later date. The similarities far outweighed the differences.

Commercial Centers Team. In the metropolitan government simulation, the allocation of larger areas of residential development in smaller groupings and lesser degree of concentration around towns demanded somewhat fewer and

presumably larger centers. The projection of the existing situation had a generally similar spatial distribution of centers and was similar in many other respects.

Transportation Team. The model worked relatively well in the first simulation, because it was able to deal with incremental changes in the location system demanded by small residential developments. However, we had difficulty with the larger allocations of industry and residence in the second run, because we were not able to handle jumps in road levels in our updating process. We started to get strong nodes, and the model did not respond too well to that type of development.

Steinitz. We will skip the political and financial groups since their work was assumed out of the second simulation.

Visual Team. In the process of doing our research we found that there are two types of landscape that can best absorb development. One is the natural landscape of the area, in this case a heavily rolling landscape; the other is an existing, built-up development. In other words, new housing that is attached to an existing pattern of development is usually quite readily absorbed. In the two simulations, development was best absorbed in the fashion you would expect: the spread development of the first by putting it into the areas of highest natural landscape absorption and the nodal development of the second by self-absorption into the man-made development. It is difficult to say which one is visually preferable.

Pollution Team. There were more economies of scale in the second run; therefore it was possible to have several residential developments of high and medium density and some industrial areas that were not allowed in the first simulation run because of a lack of sewer facilities.

---

[9]*Steinitz.* The residential model only began to work satisfactorily in the second simulation, so that residential allocations on the first are simply not as well substantiated. All these models developed as they went, and one would never in good conscience publish them without bringing out their limitations in a big heavy footnote.

Discussion

Steinitz. We are now open to questions and comments.

Rogers. First of all, I hope that you all now realize how difficult it is to write a computer program for a simulation model of urban development. Articulating as much as we have, this presentation has demonstrated that there are some very soft spots in the analysis and also some very hard spots. I think it's going to be a long time before we can get a fully programmed simulation model of this kind. That's a pity, because one of the advantages of having the model as a set of punched cards is that you can test the model rapidly. This model is full of assumptions, and we'd like to evaluate their effects on the validity of the model. For instance, in a computer-programmed model, one could very quickly test the effect of those weighting coefficients in the political model, instead of spending three months to find them out. There may be a large number of assumptions which can vary widely without making a significant difference to the outcome of the model. If we were able to get a good sense of the response of the total model to these assumptions, we would feel a lot better about using the individual parts. This applies to the whole range of models. All the assumptions you've been hearing about could be checked out in groups or one by one, to see how sensitive the system is to them. In this way we could firm-up the whole process. You can see that it took us a long time to observe the effect of just one other assumption, and a very gross assumption at that. As you have seen, the problems of sensitivity analysis on a hand-programmed simulation model are enormous.

We talked earlier about these repetitive "black boxes"—the inner workings of the models. All the parts are interchangeable. I think that we might adjust some and remove others on the basis of what we've learned in this past semester. It may be particularly helpful to use analytical methods (such as linear programming) as black boxes inside the system for some of the location problems. We have hesitated to do this because we would have, then, very large linear programs with thousands of variables and thousands of constraints. But I think the simulation system would be improved by using such analytical methods.

Parnass. I would like to say that it is not a pity this has been carried on by hand. I think that a certain amount of this hand manipulation is of value to the participants, even if we could program the whole thing for a computer.

I also think that the most beneficial thing to us was the resolution of conflicts. In the first run of Simulation One it took us almost two days to resolve the conflicts. Now (and it's almost a bad thing and I don't know how you get around it), we've become so sophisticated in terms of knowing where the conflicts are going to arise that we avoid them altogether. We already know where the bad pollution places are; we know where we're going to get the flags from the political team, so we don't allocate things there to begin with. We take the easy path. I think two more iterations on these runs and you wouldn't find a conflict at all. We're becoming too smart about this, and in this sense the simulation may even lose its pedagogic implications.

Steinitz. That's also a very dangerous thing to us as human beings. There is the danger of the program taking over. There is a very thin line to be drawn there, but at this time in my professional life I would push for a clearly structured process.

Vint. I think it's time that we talked about this process as a planning tool, which we haven't talked about much during the course. I think that if this is to be of value, it's going to have to get into a format where it is a planning tool and not simply an observational method.

Rogers. A simulation model is like a TV set with, say, 70 knobs. You turn the knobs, and the response comes out, and you quit when you have a satisfactory picture. This is the way I would view a simulation model, even in a planning context. Assume that we articulate the whole thing, and are very confident of all the component models. Then we can take it into the planning context. If you don't like the outcome, you twiddle with the knobs until you are satisfied.

Steinitz. The gadget that's operating this, be it a man or a machine, obviously has the benefit of knowledge; knowing what's going to happen through simulation, he can then come back and test six plans and use this simulation process to begin to understand the pressures that might develop if a proposed policy were to be established. An important side issue for planners is: If we foresee what happens, then what?

Breuning. There's another point in the simulation that you can't, or shouldn't, forget. You not only vary the input and look at what happens to the output, you get also the almost infinite variety of possible changes within the simulation, the changes in the policies that you are modeling.

Question. Doesn't this kind of simulation have meaning mainly when you have lots of room to shift around? Once you get into the area of an urban center, with limited land available, then it's a different problem.

Steinitz. If we can carry on next spring, we'll try just this. Including downtown Boston would be really interesting. The problem would obviously be that there is no excess of land, and we would have to solve conflicts in a much more complex way.

The kinds of solutions are going to be different, obviously. The number of conflicts are going to increase, also obviously. The solutions are going to be more complex, also obviously, and a lot of the costs borne will be politically more difficult. But the method of approaching those solutions could well be identical downtown or in the suburbs. We want to try it.

Breuning. One thing that baffles me is that you say land is available here, but not in Boston. Land is no longer available on this earth any place; it comes at a price which is measured both in dollar and political constraints. True, you may have less dollars and less political constraints out here than you have in Boston, but if you really had land available, then you neglected an important constraint. All you have to do is vary the analysis a little bit, and you can do this for downtown Boston.

Steinitz. Let's have some comments from the veterans of this exercise. It was run democratically, but the faculty came in with a pretty strong framework into which you had to put the parts. Now let's assume that we wanted to convince the chairman of the department to continue for another year. What should we tell him?

White. My feeling was that we didn't have enough work to do. We were told that this was going to be two afternoons and one evening a week, and I don't think most of us believed it because we had never had a course like this. But I feel that it would have been worthwhile to have more hours and more work and carry it farther. The fact is that all of us were essentially not getting good results from our models until toward the end, so it was very hard to know whether our good results were in fact good; and it would have been nice to have been able to carry it a little further. I think that could have been done with a little more intensified course.

Steinitz. A strange comment, but OK. In the beginning, during the development of the models, we were working three afternoons and one night a week. Then we got so efficient at the procedure, that people were in here on Monday and Wednesday at 2:00 and out at 5:00 and doing updating and other odd tasks on occasional other days. We could have done more.

Comment. What the course has done is expose me to a broad range of problems in a very detailed way—that I would not have been exposed to perhaps in another format. And I think that this method provides the opportunity to become really aware of the realities of how the world is.

Comment. One other thing should be mentioned here, and this is the idea of design trade-off. We found that there were many instances where there were many demands beginning to cluster: a shopping center, a factory, and a housing development, all demanding proximity to perhaps a major road intersection with limited land. At that point, the trade-off consisted of an immediate sketch problem—an instantaneous sketch that happens right there on the wall—in which we could try to find a way to combine all these types in a limited area. If necessary, we may even come up with a new type of construction or other design innovations, to resolve these conflicts so that nobody would have to move and everybody could satisfy their allocation priorities.

Breuning. Let me ask a question of the students. I wonder, in looking at this exercise—in putting in your decisions and in getting a feedback—what sense of reality you have? Is this what's actually going on? Assuming this condition were to remain, which is very likely in this state, do you really believe that the results of the first simulation that you did would look like this in 1995? Or do you feel that this is just a nice exercise but it could take any of 100 different shapes and forms?

Comment. I think that most of us have come to the conclusion that the general patterns are probably valid. We couldn't hope to predict the individual locations; conditions are going to change, and a small change somewhere that we haven't thought about could make a major impact on a small area. So I think only the basic pattern can be defended.

Steinitz. The development's not as scattered as you might think. It's an hour's drive maximum from one end to the other, and maybe a half hour across. I think it very reasonable. It surely could have happened a thousand different ways, but this is as likely as anything. We ended with a rather typical diagram. You'd expect the industrial development and the commercial centers to be on the highways, the residential off to the sides, and the recreation in the middle—that is how you would sum it up. But the trouble is that most people would begin that way and then not be able to go any further. The fact that we may have an output that is similar to what you'd expect is okay.

Breuning. On the contrary, it is very good. You would worry if it weren't what you expected.

Comment. If we had been given a problem statement at the beginning of the semester to develop two alternative schemes for 600,000 people coming into this region, with so much industry, etc., etc., I'm sure none of us would have gotten very far. Now I think we could take teams and develop plan alternatives based on desirable adjustments of the projective simulation.

Von Moltke. I think this is a very good point. I'd like to see the next step; to compare different plans and really evaluate them critically. Which is better from a local point of view, economically, socially, etc.?

Steinitz. One important thing is that we actually have four measures of the performance of this region over time: fiscal, political, visual, and ecological. You certainly can take the knowledge that you've gained here and come up with a plan, and subject that plan, or let's say the expectation that that plan would be built, to those four measures; see if it in fact is better than what might happen without your efforts.

Comment. It would be very interesting to see what would happen with completely different assumptions. For example, if the federal government would guarantee a certain level of education, it would change the picture of this area entirely. What effect would this have?

Steinitz. You can't do everything. We went through an afternoon meeting in which we listed a half-dozen issues, and we simply voted for the one that we wanted to do most—which was metropolitan government. A social revolution and its implications could also be simulated, as could a financial crisis or almost any other.

Breuning. I asked earlier what the class was thinking because I was somewhat surprised about your insistence on remaining involved in the model operation. I've been involved in engineering, and in a way there are similar conflicts. For instance, in highway development, the analysis of the traffic requirements, the design of the road, the testing of the road against expected traffic requirements, and the allocation of funds, all must be compromised to set an optimal improvement of the transportation system. The typical student reaction is to go on to the next set of questions, that is, what are the higher-level implications? So, I'm wondering if maybe we are begging the question here. Maybe we should pursue both approaches —develop a model which is fully automated to test the variations in input-output policy, and another one in which you can test out the internal operations of the model itself. This involves two separate approaches.

Rogers. It's very, very easy to remain sitting around and talking about the policies—without any way of handling it. If we had brought in, or had been in a position to bring in, students from political science, and sociology, etc., then I think that your idea would be more useful. We could have concentrated more effort on policy content.

Breuning. I think that I'm in full agreement with you, but I'm trying to say, let's develop the model in such a way that you can in one case leave the input-output fixed and see what happens if you change the internal analysis procedures in the model, and then in the other case, leave the model intact and see what happens when you change input-output relations.

Rogers. Just to reinforce this point, because I think it is the same point that I was trying to make earlier, the fact is that we do have a model. We know the limitations of our expertise —and it is limited. Let's assume, however, as we've done throughout the whole process, that what we have is accurate and that we can and are going to use it. Then we've learned a tremendous amount about the trade-offs between a visual evaluation and a demand of the residential developer, and of the developer and pollution, etc. Let's find out what the different consequences are of not only metropolitan government but of even more complex programs of rapid transit or of New Towns. I think that the trade-offs at this larger scale are going to be just as meaningful to us, not so much as a planning process as in getting to be more sophisticated about what the implications are of large-scale policies on small-scale trade-off situations.

Question. Was it more difficult for them to resolve conflicts between medium-cost housing and visual and pollution than it was in the cases of the other two housing-cost types?

Steinitz. Probably. The low-cost housing typically lost; the high-cost housing always won; it was the middle that was the problem and where there were often long arguments.

Question. What in fact are the priorities here? Is it right to start with industrial location, then housing, and then transportation? What would happen if you upset the whole order and took as a priority a social development or housing?

Rogers. There is an order of who goes first in this simulation model. It is a negotiated order which we decided seemed best to describe the reality. In this case, we have a time iteration of five years, which is very gross, and the order is important. Let's assume an iteration of five weeks; then it wouldn't matter very much who went first. The importance of the order is a function of the time of the iteration of the simulation model.

Answer. In the beginning of the semester, we did talk quite a bit about which would be the priming models. We looked at what's going on in the region now, and what is happening is that the industry is moving in and housing has generally followed the industry. We therefore took this as an assumption. But we could have started with any other and probably would have gotten a different configuration but, as Peter said, not a very different one.

Steinitz. With the knowledge beforehand that we only had time to do two or three runs, I think the most instructive assumption was to simulate what's going on as closely as we could interpret it.

There's another aspect to your question which is far more serious and which we're only now studying ourselves. That is, what happens in a conflict with, for example, two of the models in favor of one side, and two of the models on another? You may have a situation where your fiscal policy says "yes," your visual policy says "no," and there's no common denominator. This is a far more difficult question, one which we've talked about, for which we have no solution. I have my opinions and I think I would go to the political model to decide.

This is a far more difficult and deep line of questioning for this kind of process. We are evaluating in terms of dissolved oxygen, visual absorption, tax rates, and political liking, and those are four types of values without a common unit of measurement.

Question. Do you have any idea of how reliable each location is with respect to data? For example, it seems obvious that as

you get closer to the edge of the study zone, there is less confidence because of what is happening outside the zone.

Steinitz. The first and most obvious thing is that we can only talk about the complete, whole towns in this area, which eliminates most of the edge portions. In the specific case of transportation, we have no through-traffic capability. We don't know, for example, in an industrial location on the edge, whether the work force and its residential areas are outside. There are many cases like this, because the border of the study area is, after all, arbitrarily chosen.

I think that if we were setting up the area again, we would begin to think much more about setting an inner boundary in which we have more confidence. On the other hand, the area that we have the most confidence in is, of course, the area that interests us the most, that is, the triangle which includes the Charles River basin.

Comment. What interests me most in this approach of planning is the future; in time you can see how your work predicted a real state.

Rogers. Obviously the thing to do is to go back and look at the historical data, and to start off in 1920 and bring it up to 1965 or 1968, and see how our model predictions compare to what has happened. This is a good test of a simulation model. It is obviously very difficult to experiment in a future time.

Question. In a case where a cell was developed at a late stage, how do you protect it earlier, on the guess that in 10 years it might become valuable?

Steinitz. The way that we've been doing this, there's no reason to protect it at an early stage. There would be if you had a future view, and then retreated into the past to protect a new site or expansion. But let's not forget one thing, that in each stage, these allocation interests were actually doing the best they could. They got what they wanted, subject to the constraints of the evaluation.

Rogers. However, there was no provision within any of the allocation models that recognized the fact that an industry is more likely to locate in a certain area after three or four industries are there because all the services and the general provisions for industrial use are located there already. There has to be an external input into the allocation models which recognizes existing development and economies of scale.

Question. I think this would be a very valid addition because activities always have a tendency to cluster anyway.

Steinitz. I'm sorry, but that's not at all clear. A lot of activities want to spread themselves out, and a lot of activities, if they're responsive to real estate prices, must do so. If one develops first, just that fact might raise adjacent land prices so much that the next one couldn't cluster. The hypothesis that everything tends to cluster is certainly not a complete generalization. It may be for some activities but not for others.

In any case, the distances here are very small. I would call a lot of these clustered even if someone else would call them spread. There's no objective standard that we have to define this.

Question. Did the group have any kind of subjective evaluation, standing back and looking at the results of your work? What was the group's feeling about it?

Answer. I don't think we spent enough time discussing whether we like it or don't like it, whether it seems reasonable or not, or whether we are pleased with the development or not; it would take, I think, a bit of time to do some detailed analysis of what these patterns would really mean, what the real effects are. Don't forget that we weren't satisfied with the results of housing allocation until the second run.

Steinitz. Some people thought that it was very scattered and argued the good points of that. There are some, obviously. Scatter means that you can fill in later, and you can mix ages and types more over a long time. But some people reacted very violently against it. They just didn't like it. They looked into the future and they didn't like what they saw.

Answer. We had all expected a lot of development to take place near the Charles River. But in fact, given the analyses that we did, say on housing, we found out that it was not desirable land for housing, mainly because the quality of drainage was so low that it would raise site costs to a prohibitive extent. Our work seemed to make sense in the patterns that the large recreation areas took, and again, in the housing and industrial allocations. The over-all pattern seems to be rather convincing.

Steinitz. One very instructional thing for the landscape architecture students was that this format forced the Recreation Team to develop a strategy where priorities are stated. Put yourself in a situation using the traditional studio methods— what would you do? The first thing would be to buy the river. Right? You'd spend all your money on the river. Well, in both the simulation runs, it happened that for twenty years nobody else wanted the river. You therefore might be able to say, "Well, look, we don't have much of a problem there for a good long time. Let's go buy as much land as we can where we're going to have the problem, where it doesn't exist now, and then in the end we'll worry about the river." And in fact, that's how it turned out. And this is completely different from what you would normally expect a student plan to show with the values that the students have when they come in here. The format really makes you think of what the hell it is that you want to do and what is the best strategy for getting it.

Appendix A:
Data Inventory,
Boston Region,
Southwest Sector

The initial data collection was part of the work of an experimental program conducted during Fall 1967, in the Department of Landscape Architecture, Harvard Graduate School of Design. The course had two principal aims: (1) to introduce the use of computer analysis into the Landscape Architecture curriculum, and (2) to test several resource analysis and allocation innovations. The pedagogical emphasis was on the organization and use of data, rather than on the acquisition of the data. The students were the incoming class in Landscape Architecture, and this was their first required coursework. The caveats applicable to all student work are thus distinctly relevant to these studies.

Participants

Carl Steinitz
David Sinton, Teaching Assistant

James A. Beard
Garr Campbell
Richard Forsyth
Harry Garnham
Roger Holtman
Peter Jacobs
Knox Johnson
James Knode
William Laubmann
Richard List
Robert Longfield
Timothy Murray
Takero Ogawa
Mrs. Beatrice Pettit
Benjamin Reed
Keith Renner
Richard Rigterink
Andrew Sammataro
Charles Smith
Wayne Tlusty
Albert Veri
Douglas Way

**Data Inventory by One Kilometer Square Grid.
Boston Region, Southwest Sector**

Card	Columns	
1	2–5	Cell Identification Number
1	7	"Good" Agriculture, Percent of Area
1	8	"Poor" Agriculture, Percent of Area
1	9	Forest, Percent of Area
1	11	Residential Density
1	12	Residential Cost and Quality
1	13	Residential, Predominant Type
1	14	Residential, Age of Development
1	15	Residential, Percent of Area
1	17–18	Commercial, Number of Establishments
1	19–20	"Heavy" Industry, Number of Establishments
1	21–22	"Light" Industry, Number of Establishments
1	24	Institutions and Services, Percent of Area
1	25	Institutions and Services, Major Type
1	26	Recreation, Percent of Area
1	27	Recreation, Major Type
1	29	Recreation, Access
1	30	Road Transport, 1965, Major Road Type (incl. proposed Rt. 495)
1	31	Road Transport, 1965, Major Road Type (without Rt. 495)
1	33	Road Transport, 1965, Average Daily Car Volume
1	34	Air and Rail Transport, Major Type
1	35	Water, Percent of Area
1	36	Water, Major Type
1	38	Water Quality; Source: Mass. Water Resources Commission
1	39	Water Navigation, by Largest Craft
1	40	Topography, Slope
1	41	Drainage, Percent Well Drained
1	43	Bedrock
1	45(44)	Environmental Nuisances, Summary Values
1	48(47)	Environmental Nuisances, Major Type
1	50	Visual Texture and Landscape Variation
1	51	Visual Closure
1	52	Visual Character, Predominant Type at Most Public Area
1	53	Visual Character, Secondary Type
1	56–58	Land Cost
1	59	Land Cost (Recoded)
2	2–5	Cell Identification Number
2	7	Water Supply Potential
2	9	Sewer Facilities & Potential Capacity
2	11	Garbage Disposal Units (data only for MAPC area)
2	14–15	Distance from Elementary Schools (within towns)
2	20–21	Access to Route 128
2	22–23	Access to Providence, R.I.
2	24–25	Access to Framingham
2	26–27	Access to Interchange of 495 & 95
2	30–31	Road Transport 1968, Major Road Type (incl. Rt. 495)
2	32–33	Access to Limited Access Highways
2	34–35	Access to Limited Access Highway Interchanges
2	36–37	Vegetation Density
2	38–39	Topography, Visual Closure: Largest Percent of Cell
2	40–41	Topography, Visual Closure: Least Absorptive Part
2	42–43	Topography, Visual Closure: Most Absorptive Part
2	44–45	Degree of Visual Effect (like-dislike)
2	46–47	Degree of Complexity
2	48–49	Soil Texture
2	50–51	Travel Time by Public Transit to Downtown Boston
2	60–64	Estimation of Visual Quality
2	65–69	Topographic Elevation in Feet
2	71–72	Town Index
2	80	Card No. 2

Data Inventory by Cities and Towns
Boston Region: Southwest Sector

(All Data are from the U.S. Census, 1960, unless otherwise noted. An asterisk following the listing refers to the source. *Town and City Monographs*, Revised (Massachusetts Department of Commerce, 1964)

Card	Columns	
1	1–3	Number of Town
1	4–8	Number of State
1	9–28	Name of City or Town
1	29–32	Area (in square miles by tenths)
1	33–39	Total Population 1960
1	40–46	Total Population 1950
1	47–53	Total Population 1940
1	54–60	White Population
1	61–67	Total Number of Households
1	80	Number of Card: 1
2	1–3	Number of Geographic Unit
2	4–10	Population in Households
2	11–6	Male Population under 5 Years
2	17–22	Male Population 5 to 14 Years
2	23–28	Male Population 65 Years and Over
2	29–34	Female Population under 5 Years
2	35–40	Female Population 5 to 14 Years
2	41–46	Female Population 65 Years and Over
2	47–51	Percentage Difference in Population 1950–60
2	52–54	Percent of Population Nonwhite
2	55–58	Percent of Population under 18 Years
2	59–62	Percent of Population 18–64 Years
2	63–66	Percent of Population 65 Years and Over
2	67–70	Fertility Ratio
2	80	Number of Card: 2
3	1–3	Number of Geographic Unit
3	4–10	Population in Households
3	11–15	Percent Increase of Population in Households, 1950–1960
3	16–19	Population per Household
3	20–23	Percent of Population in Group Quarters
3	24–27	Percent of Population Foreign Born
3	28–31	Percent of Population Moved into House after 1958
3	32–35	Percent of Population Migrant
3	36–39	Median School Years Completed
3	40–43	Nonworker/Worker Ratio
3	44–47	Females 14 Years and Over (Percent in labor force)
3	48–51	Males 18–24 Years (Percent in labor force)

Card	Columns	
3	52–55	Males 65 Years and Over (Percent in labor force)
3	56–58	Percent of Civilian Labor Force Unemployed
3	59–62	Percent of Total Employed Who Are in Manufacturing
3	63–66	Percent of Total Employed Who Are in White Collar Occupations
3	67–70	Percent of Total Employed Who Are Working Outside County of Residence
3	80	Number of Card: 3
4	1–3	Number of Geographic Unit
4	4–7	Percent of Workers Using Public Transport
4	8–12	Median Income for Families in Dollars
4	13–16	Percent of Families with Incomes under $3,000
4	17–20	Percent of Families with Incomes of $10,000 or Over
4	80	Number of Card: 4
5	1–3	Acronym of Geographical Unit
5	5–7	Average Family Size*
5	9–12	Pupil-Teacher Ratio*
5	14–17	Educational Dollar Expenditure per Pupil*
5	19–22	Vehicular Trip Ends per Square Mile
5	24–26	Truck Trip Ends per Square Mile
5	28–31	Mass Transit Trip Ends per Square Mile
5	33–36	Percent of Families with Incomes $3,000–$5,000
5	38–41	Percent of Families with Incomes $6,000–$9,999
5	43–46	Percent of Employed Population in Professional and White Collar Occupations*
5	48–51	Percent of Employed Population in Skilled Labor Occupations*
5	53–56	Percent of Employed Population in Unskilled Labor Occupations*
5	58–61	Average Land Cost per Acre (1963)
5	63–66	Median House Value in Thousands*
5	68–71	Percent of Houses in Middle Value Range*
5	80	Number of Card: 5
6	1–3	Acronym of Geographical Unit
6	5–8	Average January Temperature in Degrees F.*
6	10–13	Average July Temperature in Degrees F.*
6	15–18	Average Yearly Precipitation in Inches*
6	20–23	Median Monthly House Rent in Dollars, Including Utilities*

Data Inventory by Cities and Towns

Card	Columns	
6	40–43	Percent of Registered Voters Recorded as Republicans, 1966
6	72–74	Vertical Distance of Town Centroid from 0, 0; in Inches
6	76–78	Horizontal Distance of Town Centroid from Point 0, 0; in Inches
6	80	Number of Card: 6

Card 7 is 1950 data ⎫
Card 8 is 1960 data ⎬ as listed below
Card 9 is 1965 data ⎭

Municipal Expenditures
(Data are expenditures per capita, rounded to 1/10 dollar)

Columns	
1–3	Abbreviation of Town Name
4–9	Population*
10–15	Total Expenditures
16–20	General Government Expenditures
21–25	Public Safety Expenditures
26–30	Health and Sanitation Expenditures
31–35	Highway Expenditures
36–40	Public Assistance Expenditures
41–45	Veterans' Services Expenditures
46–50	School Expenditures
51–55	Library Expenditures
56–60	Recreation Expenditures
61–65	Pension Expenditures
66–70	Unclassified Expenditures
71–75	Public Service Enterprise Expenditures
76–78	*If no data were available for correct year, or years immediately following or preceding, this field indicates the abbreviation of the town whose data were substituted. Alternatively, if data were not available for correct year but were available for year following or preceding, the columns indicate the substituted year.
80	Data Card No.

Card	Columns	
10	6–15	1965 Tax Base
10	18–20	1965 Tax Rate
10	22–25	1965: Number of New Family Accommodations as represented by building permits issued for new construction, not including apartments created through improvements to existing buildings and by assessor's estimates where permits are not issued.

Card	Columns	
10	27–30	1966: Same as above
10	31–35	Number of Public Housing Units sponsored by either the State of Massachusetts or the Federal Government. In some instances these figures have been reduced by sale of houses. They represent only permanent units undertaken between January 1, 1946 and June 30, 1967, reported by the Division of Housing. In a very general way these figures, deducted from the totals of building permits issued by the respective towns, give an indication of the number of units privately undertaken. Such a calculation, however, should not be regarded as precise measure, since in some instances permits were issued for temporary and other public housing units that are not included in the figures.
10	36–40	1965 Density of Persons per Square Mile (based on land area only)
10	41–45	% of Population Change 1965 over 1955
10	47–50	Year Zoning Ordinance Enacted
10	51–55	Minimum Area for Single-Family Residential Use (square feet)
10	58–60	Minimum Frontage of Single-Family Residential Lot (linear feet)
10	61–65	Minimum Area for Commercial or Industrial Use (square feet)
10	68–70	Minimum Frontage for Commercial or Industral Use (linear feet)

Appendix B:
The Grid Program

GRID is a computer program created by David Sinton and Carl Steinitz at the Laboratory for Computer Graphics, Harvard University, specifically to provide a highly efficient means for graphic display of large quantities of information collected on the basis of a rectangular coordinate grid (Figures B.1–B.5). The program is written in FORTRAN IV and is currently being operated on an IBM 7094 with a 32k memory, or an IBM 360/65, using 120k byte memory. With some internal adjustments, it can be run on a computer with a memory as small as 12k words. The program requires two sets of data input—first, the data values associated with a spatial grid, and second, a series of instructions to the program about the particular procedures and forms that are to be used for analysis and display.

Each data value is assumed to be associated with a cell on the grid. It is essential that the values be processed in the correct order, since the program accepts the data in the order in which it prints the maps. By the standardized printing process, the program starts at the top of the map and processes the data horizontally, row by row, and from left to right in each row. The numbers below represent the order in which 30 data values in a six-by-five grid will be printed and processed.

1	2	3	4	5	6
7	8	9	10	11	12
13	14	15	16	17	18
19	20	21	22	23	24
25	26	27	28	29	30

The user specifies the size and the shape of his grid. While the program is normally used in rectangular grids, it provides two methods of specifying irregular outlines. The program has been designed with an internal loop that permits an unlimited number of cells to be mapped. However, in normal usage, it is not expected that the average grid will be greater than 10,000 data points.

Before printing the spatial diagram, the actual data values are generalized into groups, each group having a unique graphic symbol associated with it. Using options, the user can then specify the number of levels, the maximum value of the data, the minimum value of the data, and the relative size of each of the levels in the range of the data. Thus, the user has complete control over the levels into which his data are divided. The user also has control over the symbolism which is used to print the spatial diagram, e.g., a gray scale between white and black, a dot map, or any alphanumeric symbols. The program will also print specified information about the data analyses which are being mapped, and it will print the numbers of the grid coordinate system around the edges of the map.

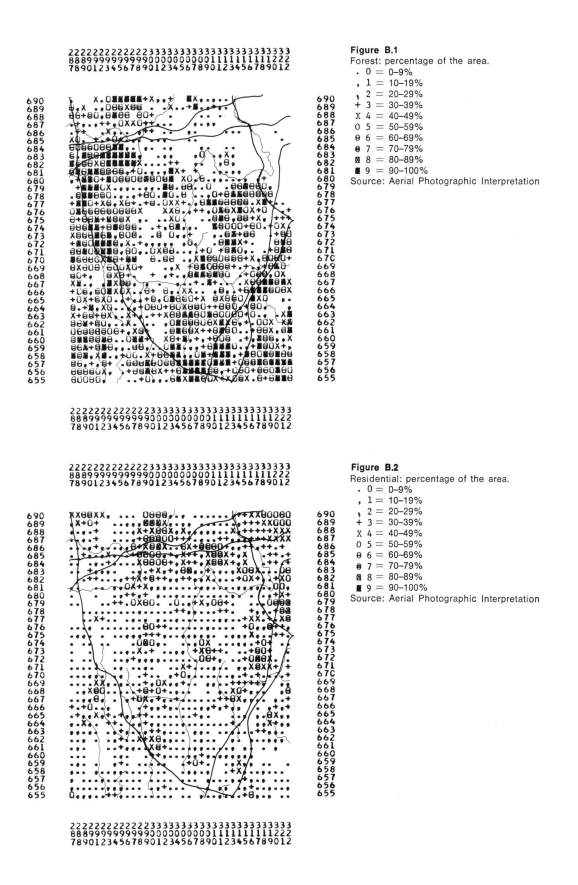

Figure B.1
Forest: percentage of the area.

. 0 = 0–9%
, 1 = 10–19%
، 2 = 20–29%
+ 3 = 30–39%
X 4 = 40–49%
0 5 = 50–59%
θ 6 = 60–69%
θ 7 = 70–79%
⊠ 8 = 80–89%
■ 9 = 90–100%

Source: Aerial Photographic Interpretation

Figure B.2
Residential: percentage of the area.

. 0 = 0–9%
, 1 = 10–19%
، 2 = 20–29%
+ 3 = 30–39%
X 4 = 40–49%
0 5 = 50–59%
θ 6 = 60–69%
θ 7 = 70–79%
⊠ 8 = 80–89%
■ 9 = 90–100%

Source: Aerial Photographic Interpretation

Figure B.3

Average land cost per acre, rescaled.

. 0 = Public Land
, 1 = $100–$500
, 2 = $500–$800
+ 3 = $800–1,000
X 4 = $1,000–1,200
0 5 = $1,200–1,500
θ 6 = $1,500–1,800
θ 7 = $1,800–2,500
⊠ 8 = $2,500–3,000
■ 9 = $3,000 and over

Source: R. M. Bradley and Company, Boston

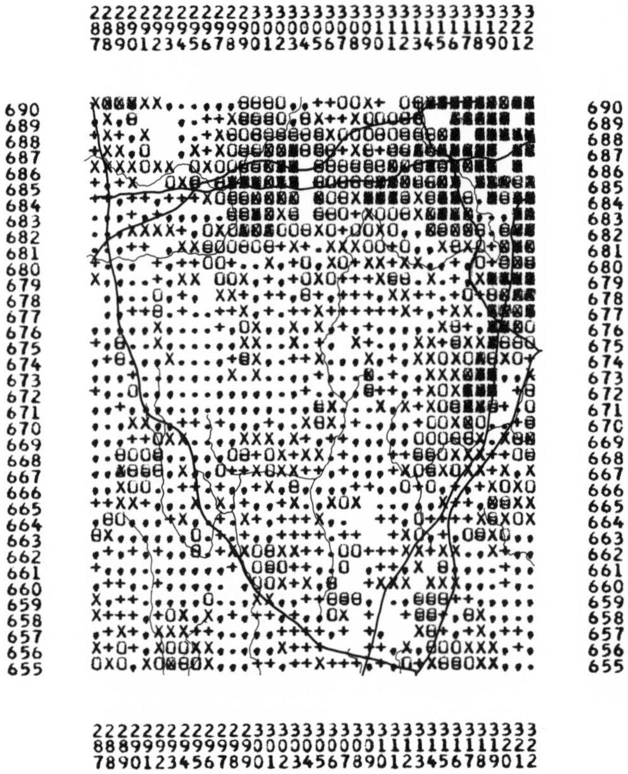

Figure B.4

Topography: elevation.

Elevation is mapped in ten equal levels between minimum and maximum elevations.

. 0 = 40′–91′
, 1 = 92′–142′
, 2 = 143′–193′
+ 3 = 194′–244′
X 4 = 245′–295′
0 5 = 296′–346′
θ 6 = 347′–397′
θ 7 = 398′–448′
⊠ 8 = 449′–499′
■ 9 = 500′–550′

Source: U.S.G.S. Map

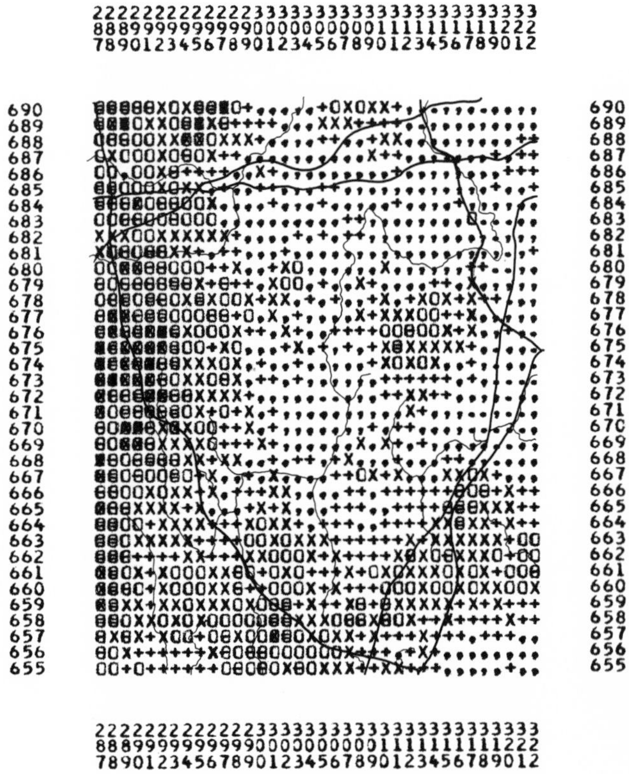

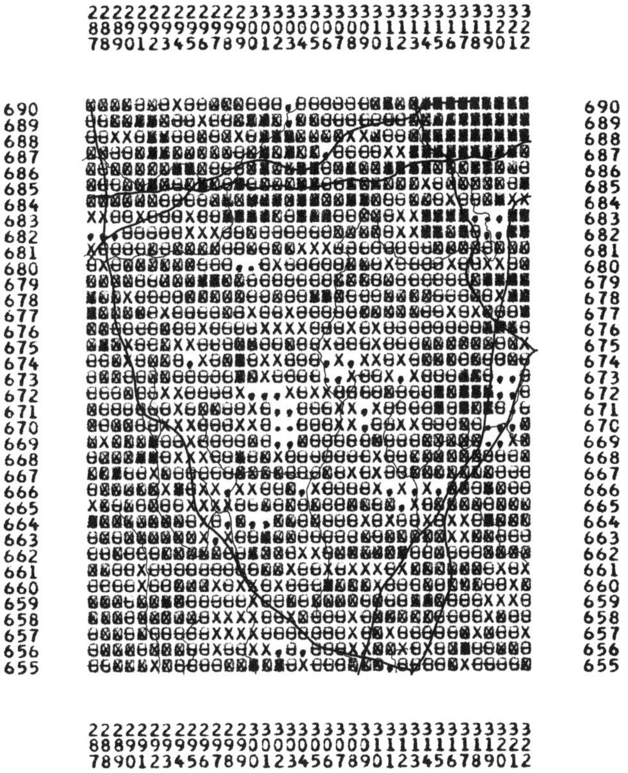

Figure B.5
Good drainage: percentage of area.
- · 0 = 0–19%
- ₁ 2 = 20–39%
- X 4 = 40–59%
- θ 6 = 60–79%
- ℕ 8 = 80–99%
- ■ 9 = 100%

Source: Aerial Photographic Interpretation

Selected References

Aguilar, Rodolfo J., and James E. Hand. "A Generalized Linear Model for Optimization of Architectural Planning." Submitted to Joint Computer Conference, Spring 1968.

Boulding, Kenneth. "The Economist and the Engineer: Economic Dynamics of Water Resource Development," in *Economics and Public Policy in Water Resource Development*. S. Smith and E. Castle (eds.), 1964.

Bulkley, Jonathan W., and Ronald T. McLaughlin. "Simulation of Political Interaction in Multiple-Purpose River-Basin Development," M.I.T. Hydrodynamics Laboratory Report No. 100, October 1966.

Donnelly, Thomas G., F. Stuart Chapin, Jr., and Shirley F. Weiss. "A Probabilistic Model for Residential Growth." Institute for Research in Social Science, University of North Carolina, Chapel Hill, 1964.

Feldt, Allan G. "Operational Gaming in Planning and Architecture." AIA Architects-Researchers Conference, Gatlinburg, Tenn., October 1967.

Harrington, Joseph J. "Operation Research—Relatively New Approach to Managing Man's Environment." *New England Journal of Medicine*, Vol. 275 (1966).

Harris, Britton. "Organizing the Use of Models in Metropolitan Planning." Seminar on Metropolitan Land Use Models, Berkeley, Cal., March 1965.

———. "The Uses of Theory in the Simulation of Urban Phenomena." Presented at Highway Research Board meetings, Washington, D.C., January 1966.

Isard, Walter, and Robert E. Coughlin. *Municipal Costs and Revenues Resulting from Community Growth*. Wellesley, Mass.: Chandler Davis Publishing Co., 1957.

Jacobs, Peter, and Douglas Way. "Visual Analysis of Landscape Development." Monograph, Department of Landscape Architecture, Harvard Graduate School of Design, Cambridge, Massachusetts, 1968.

Kilbridge, M., R. O'Block, and P. Teplitz. "A Conceptual Framework for Urban Planning Models." *Management Science*, Vol. 15, No. 6 (1969).

Lamanna, Richard A. "Value Consensus Among Urban Residents." *Journal of the American Institute of Planners*, Vol. 30, No. 4 (1964).

Lowry, Ira S. *A Model of Metropolis*. Santa Monica, Cal.: The RAND Corporation, 1964.

Maass, Arthur. "Benefit-Cost Analysis: Its Relevance to Public Investment Decision." *Quarterly Journal of Economics*, Vol. 80 (1966).

———. "A Short Course in Model Design." *Journal of the American Institute of Planners*, Vol. 31, No. 2 (1965).

Morrill, Richard L. "The Negro Ghetto: Problems and Alternatives." *Geographical Review*, Vol. 55 (1965).

Morse, Philip M. (ed.). *Operations Research for Public Systems*. Cambridge, Mass.: The M.I.T. Press, 1967.

Pool, Ithiel de Sola. "Simulating Social Systems." *International Science and Technology*, March 1964.

Simon, Herbert A. *The New Science of Management Decision*. New York: Harper and Row, 1960.

Thomas, Harold A., Jr. "The Animal Farm: A Mathematical Model for the Discussion of Social Standards for Control of the Environment." *Quarterly Journal of Economics*, Vol. 77 (1963).

———. "Use of Mathematics in Planning." Mimeo. Harvard University, Cambridge, Mass., May 1961.

Tri-County Regional Planning Commission, Lansing, Michigan. "Models for Predicting Employment, Population, and Land Use (Exclusive of Transportation Networks)." Mimeo. Prepared by Community Systems Foundation, Ann Arbor, Michigan, December 1964.

Weathersby, George. "Quantitative Urban Analysis—A Synthesis of Decision Theory and Modern Control Theory." Mimeo. Harvard University, Division of Engineering and Applied Physics, Cambridge, Mass., Spring 1968.

Wilson, James Q. *The Metropolitan Enigma: Inquiries into the Nature and Dimensions of America's "Urban Crisis."* Washington, D.C.: U.S. Chamber of Commerce, 1967.

———, and Edward Banfield. "Voting Behavior on Municipal Public Expenditures: A Study in Rationality and Self-Interest," in Julius Margolis (ed.), *The Public Economy of Urban Communities*, Washington, D.C.: Resources for the Future, Inc., 1964.